Regulating International Trade in Wildlife

The Convention on International Trade in Endangered Species of Wild Fauna and Flora (CITES) was one of the first in a new wave of global multilateral environmental agreements (MEAs) formed after the 1972 United Nations Conference on the Human Environment. It is widely recognized as being one of the most successful biodiversity MEAs.

Regulating International Trade in Wildlife: 50 Years of the Convention on International Trade in Endangered Species of Wild Fauna and Flora represents the first published history of the Convention to coincide with the 50th anniversary of its entry into force. It examines the context under which the Convention was created and charts the development of its internal organization, including its governing and intersessional bodies and its secretariat, and the personalities that built them.

CITES' business is about the conservation of species affected by international trade. This book reviews the application of CITES to whales, the African elephant, crocodilians, vicuña, timber and tree species, fish and other marine species. The Convention's compliance framework is arguably its most pre-eminent feature. The book looks at the way that key obligations on Parties have been identified, the standards of implementation set, what efforts are made to help Parties comply with these, and what sanctions are taken against Parties that fail to implement them.

Finally, the author offers some personal reflections on the successes and failures of CITES, based on his 40 years of engagement with the Convention, and looks ahead to how it might develop in its second half-century. *Regulating International Trade in Wildlife: 50 Years of the Convention on International Trade in Endangered Species of Wild Fauna and Flora* is of great interest to wildlife scientists, including conservationists, historians, biologists, as well as environmental scientists and non-governmental organizations.

Features:

- Tracks the history of the development of CITES, its forerunners, origins, negotiation, and entry into force
- Reviews the internal development of the Convention and its place in the wider debate about environmental security
- Is extensively referenced from primary research of the Convention's archives
- Offers incisive analysis of its successes and failures

Regulating International Trade in Wildlife

50 Years of the Convention on International Trade in Endangered Species of Wild Fauna and Flora

David H.W. Morgan

CRC Press
Taylor & Francis Group
Boca Raton London New York

CRC Press is an imprint of the
Taylor & Francis Group, an **informa** business

Designed cover image: David H.W. Morgan

Credit: Photo of press conference on seizure of pangolin scales, Elizabeth John/TRAFFIC.

First edition published 2026
by CRC Press
2385 NW Executive Center Drive, Suite 320, Boca Raton FL 33431

and by CRC Press
4 Park Square, Milton Park, Abingdon, Oxon, OX14 4RN

CRC Press is an imprint of Taylor & Francis Group, LLC

© 2026 Taylor & Francis Group, LLC

ISBN: 978-1-032-89336-5 (hbk)
ISBN: 978-1-032-86638-3 (pbk)
ISBN: 978-1-003-54227-8 (ebk)

DOI: 10.1201/9781003542278

Typeset in Times
by Apex CoVantage, LLC

This book is dedicated to those who placed their confidence in me at various stages of my CITES journey: Dr Michael J. Ford and the late A.J.B. (John) Rudge at the UK CITES Scientific Authority for animals, who introduced me to the Convention; Robin Sharp, at the UK government's Department of the Environment, who gave me a chance to work on the subject internationally; Willem Wijnstekers, in the European Commission and the CITES Secretariat, from whom I learnt so much about the detail of the Convention; and John Scanlon at the Secretariat who persuaded me to convert from CITES scientist to conference organizer. For 40 years, my wife, Françoise, has lived and breathed CITES almost as much as I have, and her wise counsel and endless encouragement have allowed me to do things which I would never have achieved on my own. Thank you.

Contents

Acknowledgements

The world of CITES is full of extraordinary people, many of whom have indirectly contributed to the book. I do, however, owe a particular debt of gratitude to Jonathan Barzdo, Dan Challender, Hélène Gandois, Charlie Manolis, John Scanlon, and Grahame Webb for their help in putting it together. I also thank Steve Ellis and M-J. Morgan for their kind assistance. The book, however, has been largely a personal odyssey, and so errors and omissions are all my own work.

Preface

In the 1960s, awareness about the fragility of the world's natural environment began to grow. The increasingly numerous and prosperous human population was having a negative impact on the conservation status of wild animals and plants. One of the reasons for this was the international trade in these species. In an increasingly interconnected world, recognition of this impact, and the international response which was required to address it, was acted upon by a handful of far-sighted people. Their persistence led many years later to the adoption of the Convention on International Trade in Endangered Species of Wild Fauna and Flora. As one of the first of a new generation of multilateral environmental agreements (MEAs), the Convention had to find its way in developing an internal structure and a modus operandi much of which was later copied by other MEAs.

From the inventiveness of its originators, through the wisdom of its early practitioners, a vibrant and effective international tool was developed to limit the damage caused by international trade in wildlife and so help promote the sustainable use of world's natural resources upon which we all depend.

As the Convention reaches 50 years since it first came into force, this book charts its origins, its development over time, and its successes and failures. In doing so, it references some of the remarkable people who have made it what it is today. Many of them have remained involved for long periods of time, which is a testament to the Convention's appeal. There truly is a 'CITES community' and it is this community that has developed the Convention, despite the derisory resources devolved to its implementation. Most of this implementation goes on at national level amongst the government officials and network of national partners who make things happen on the ground. This work is unglamorous and largely unseen, but it is where the Convention's objectives are attained or not. It seems a little unfair therefore that this book focusses mostly on international developments within the Convention itself. Whilst I spent many years toiling on the CITES coalface myself, this was only in one Party, and my experiences cannot adequately reflect the variety and richness of efforts in over 180 countries.

The idea of writing an unofficial history of CITES for its fiftieth anniversary came to me in 2017, lying in a hammock in Enyellé in the forests of the Republic of Congo. Unfortunately, circumstances have conspired to give me less time than I had hoped to undertake this task. I do hope, however, that this modest effort with all its flaws can do justice to the herculean efforts made by so few to try and ensure that future generations have the opportunity to benefit from the natural resources of the world in the same way that past and present generations have.

David H.W. Morgan
February 2025

Foreword

I've often wondered whether the Convention on International Trade in Endangered Species of Wild Fauna and Flora (CITES) should come with a warning label to users that it can become addictive!

Over the course of 50 years, CITES has attracted a deeply committed, extremely loyal and knowledgeable constituency, amongst Parties and their authorities, the Secretariat and an ever-growing array of observers. These people have dedicated countless hours to CITES and have made the Convention what it is today, creating a dynamic legal and policy framework to guide the thousands of officers serving in the frontlines to ensure its effective implementation.

David Morgan is one of those people who is clearly addicted to CITES! He worked on CITES implementation for the UK Government and the European Commission for over 20 years before joining the Secretariat where he stayed for 20 years, both as Chief of Scientific Services and Chief of Governing Bodies.

I was most fortunate to have David by my side in the Secretariat throughout my eight years as Secretary-General. Across six chapters, David provides invaluable personal insights into the origins, evolution, and future direction of CITES. In doing so, he reminds us of the key people who have driven this change; recognising that just about everyone who is closely involved in CITES can point to a decision, resolution, or listing they have influenced!

This insightful book is written by someone with deep personal knowledge of, and unique insights into, CITES. It does justice to the extraordinary efforts that have been made over the past six decades to advocate for, create, and continually improve, this remarkable global agreement to ensure our planet's extraordinary wildlife is not overexploited through international trade.

On its 50th anniversary of entry into force, it's also timely to pause and pay tribute to the negotiators of CITES for concluding a convention that is focussed and pragmatic and structured to enable it to evolve over time. Considering the scale of the decline of wildlife populations, and the links to healthy ecosystems and climate change, CITES is a convention that is needed as much today, if not more, than when it was first contemplated by those creative and forward-looking people in the 1960s and 1970s.

In answer to its critics, yes, CITES is imperfect and must further improve its effectiveness, but because of its sound structure, coupled with Parties' ability and willingness to adapt to change, it has stood the test of time. In fact, for reasons well-articulated by David, CITES stands out amongst the 1,400 multilateral environmental agreements adopted since the 1972 Stockholm Conference on the Human Environment.

This book is a riveting read, and I am sure that old hands and novices alike will thoroughly enjoy learning from David about the unique history of the genesis, adoption, and implementation of this truly remarkable and addictive convention!

John E Scanlon AO
Executive President, International Council of Environmental Law
Trustee, Royal Botanic Gardens, Kew
Chair, UK Illegal Wildlife Trade Challenge Fund
CITES Secretary-General, 2010–2018

Author

David H.W. Morgan worked with the Convention on International Trade in Endangered Species of Wild Fauna and Flora (CITES) in various capacities for 40 years. He spent 17 years as a nature conservation adviser to the Government of the United Kingdom and 6 years in the environment directorate of the European Commission in Brussels, Belgium. He worked at the CITES Secretariat in Geneva, Switzerland from 2004–2023, for most of the time as Chief of Scientific Services, but also as Chief of Governing Bodies and nine months as Officer-in-Charge of the Secretariat. He is an elected member of the UK's Institute of Ecology and Environmental Management and a Chartered Environmentalist with the Society for the Environment.

Referencing of Official CITES Documents

Official documents of the meetings of the Conference of the Parties (CoP), produced by the CITES Secretariat, are cited thus:

Resolution Conf.—Resolutions of the Conference of the Parties.

Dec.—Decisions of the Conference of the Parties.

Doc.—Working documents of the Conference of the Parties or intersessional committees produced in advance of the meetings.

Inf.—Information documents of the Conference of the Parties or intersessional committees.

Rep., Rec. or Summary record—official records of a meeting or session of the CoP.

Prop.—Supporting statements for proposals to amend the Appendices put to meetings of the CoP.

Com., Plen., Com. I or Com. II—in session documents of the CoP or intersessional committees.

As a general rule, the first Arabic number refers to the number of the CoP or inter-sessional committee meeting, and the subsequent Arabic numbers are the document serial numbers.

Up to and including CoP11 in 2000 the results of its meeting were published in paper form as proceedings of the meeting concerned. For some time thereafter, they were distributed to Parties in paper form with Notifications to the Parties and published electronically on the CITES website, before the former practice was discontinued.

1 Genesis and Principal Features of the Convention

Humans *Homo sapiens* are but one of the estimated nine million species on earth, and like all the others, they do not exist in isolation. They depend, directly and indirectly, on other species around them for their existence. Prior to the development of agriculture, humans were hunter-gatherers and depended on wildlife—wild animals and plants—for their survival and well-being. Today, billions of people around the world still rely on wildlife for food, energy, construction materials, medicine, and other uses, 70% of the world's poor directly dependent on wild species.[1]

Barter and trade of wildlife and wildlife products has doubtless always occurred, but the first evidence of international trade in wildlife probably dates from around 12,000 years ago, involving trade in spices such as cinnamon *Cinnamomum* spp. and black pepper *Piper nigrum* from south Asia to Europe.[2] Around 4–5,000 years ago, the ancient Egyptians were using and trading ivory, animal skins, and live animals for menageries.[3] For millennia, low human populations with what would be considered inefficient harvesting techniques today meant that the impact of this trade was low, and international trade in wildlife continued unchecked and with little or no limit, other than traditional or customary ones established by peoples who lived with and used the species concerned. However, the period of overseas exploration by sailors from a number of European countries from around the fifteenth century and the development of colonization and international trade routes began to lead to overuse of species for international trade. The range of uses is huge: live specimens for pet animals or ornamental plants and parts and derivatives for leather and garment fashion items, food ingredients, construction and other household goods, musical instruments, tourist curios, fragrances and medicines. Sometimes this resulted in global extinction of the species involved.[4] Concerns about the adverse impact of trade on species began to be reflected in civil society movements in some countries in the 1800s. The Society for the Protection of Birds in the United Kingdom of Great Britain and Northern Ireland (UK), now a conservation charity with over 1 million members, began in 1889 as a pressure group opposed to international trade in feathers for the millinery industry.[5] Their pressure finally resulted in legislative action under the Importation of Plumage (Prohibition) Act in 1921. In the United States of America (USA), the 1900 Lacey Act, outlawing the interstate transportation of illegally killed animals, was amended in 1935 to make it unlawful to import wild mammals and birds taken or exported illegally from abroad.[6]

DOI: 10.1201/9781003542278-1

1.1   THE ORIGINS OF THE CONVENTION

Early attempts to establish international agreements between countries on this issue arose from colonial concerns, particularly about the need to conserve species subject to trophy hunting. The Anglo-German 1900 London Convention Designed to Ensure the Conservation of Various Species of Wild Animals in Africa which are Useful to Man or Inoffensive, also titled the Convention for The Preservation of Wild Animals, Birds, and Fish in Africa, failed to come into effect for want of sufficient ratifications by potential member Parties. However, it formed the basis for the 1933 London Convention Relative to the Preservation of Fauna and Flora in their Natural State, which contained provisions requiring contracting Parties to prevent the import or export of hunting trophies from some 40 species unless they had been taken in accordance with national laws and regulations. As newly independent African States sought to establish their own identity, the 1933 London Convention was replaced in 1968 by the African Convention on the Conservation of Nature and Natural Resources. Amongst its provisions were two articles on traffic in wildlife specimens and trophies. The Convention applied to both animal and plant species, classified them in two categories depending on their degree of protection, and required different import and export protocols for each. When the text of the African Convention was revised in 2003 the provisions related to import and export of wildlife were deleted as they had been overtaken by CITES. Other early regional wildlife agreements also included provisions related to the international trade in wildlife. One such example was the 1940 Washington Convention on Nature Protection and Wild Life Preservation in the Western Hemisphere, where 19 contracting Parties agreed that:

ARTICLE IX

Each Contracting Government shall take the necessary measures to control and regulate the importation, exportation and transit of protected fauna and flora or any part thereof by the following means:

1. The issuing of certificates authorizing the exportation or transit of protected species of flora or fauna, or parts thereof.
2. The prohibition of the importation of any species of fauna or flora or any part thereof protected by the country of origin unless accompanied by a certificate of lawful exportation as provided for in Paragraph 1 of this Article.

The International Union for Conservation of Nature (IUCN), originally named the International Union for the Protection of Nature (IUPN), was established in 1948 and quickly became a key forum for the development of initiatives related to the conservation of nature. In a number of its triennial general assembly meetings during the 1950s and early 1960s, IUPN/IUCN members expressed concern about the impact of international trade on species. Animals exhibited in zoos, large African primates, leopard (*Panthera pardus*) skins, birds-of-paradise (*Paradisaeidae* spp.) feathers, and orangutans (*Pongo* spp.) were particularly mentioned, and resolutions to address these specific issues were adopted. Armand Sunier, director of Amsterdam Zoo and former president of the International Union of Directors of Zoological Gardens, submitted a broader draft resolution about this issue to the third General Assembly of IUPN in 1952, and after consideration by an in-session committee, the following resolution was adopted.[7]

RESOLUTION 96

It is desirable that in all countries the importation of animals belonging to species which are protected in their natural habitat should be prohibited, unless it has been definitely established that the exportation of such animals from their country of origin has been carried out under completely legal conditions. In those countries where such legislation is already enacted, it is highly desirable that it should be rigorously enforced.

The main focus here was on process, increasing controls on importation and ensuring legality of export, rather than on the impact trade was having on the conservation of the species—although these are clearly linked. A similar but more elaborate text was adopted at the seventh General Assembly of IUCN in 1960[8]:

RESOLUTION 14

- believing that a major threat to the existence of some rare animals is their illegal exportation from the country of origin, followed by their legal importation into other countries;
- warmly approves the action of those countries which have restricted the importation of such animals;
- now resolves that the International Union for Conservation of Nature and Natural Resources should urge all governments who do not yet restrict the importation of rare animals in harmony with the export laws of the countries of origin, to do so now and thereby support the efforts of those countries to preserve animals in danger of extermination.

The resolution was restricted to animal species, but the link between trade and the continued existence and preservation of the species was specifically mentioned.

IUCN also organized an influential symposium on Conservation of Nature and Natural Resources in modern African States, which took place in Arusha, Tanganyika (now the United Republic of Tanzania) in September 1961. At the Arusha meeting, an American ecologist Lee Talbot, then conducting a wildlife research project in East Africa on behalf of various US institutions, chaired a session on poaching and the prevention of species extinctions stimulated by international demand and trade. The consensus was that demand from Europe and the USA motivated illegal killing and that an international agreement was needed.[9]

In the spring of 1963, the chair of the newly constituted IUCN Legislation and Administration Commission, the German Wolfgang Burhenne who had studied forestry and law, persuaded IUCN's Executive Board to allow a circular to the authorities of 125 countries, requesting information on their regulation of import, export, and transit of endangered species of fauna and flora. Within IUCN, Burhenne spearheaded the idea that a multilateral agreement to address this issue was needed. According to the minutes of the IUCN Executive Board meeting of 15–17 May 1962, the initial idea was to examine the existing Convention on International Transport of Animals to see how it might be amended to include consideration of the trade in wild animals. The Council of Europe was mentioned as a possible backer for such an initiative.[10] Mention of the Council of Europe suggests that this was a reference to the European Convention for the Protection of Animals during International Transport, being discussed at the time and agreed in December 1968.

The results from IUCN's international survey were subsequently reported to the eighth IUCN General Assembly, held in Nairobi, Kenya, from 16–24 September 1963. In his role as chair of the Committee on Legislation and Administration, Burhenne reported that 'the strongest measures will prove ineffective, unless ALL Governments come to an agreement'.[11] The moment had come to propose multilateral action. The General Assembly adopted Resolution number 5 as follows:

RESOLUTION ON ILLEGAL TRAFFIC IN WILDLIFE SPECIES

Whereas many rare and vanishing species of wildlife are threatened with early extinction through illegal export from their native land and whereas such illegal export would be much less frequent if import into other countries were prohibited; recalling resolution 2,213 adopted by the General Conference of UNESCO at its 12th session and resolution 1931 (xvii) adopted by the General Assembly of the United Nations at its 17th session concerning economic development and conservation of natural resources, flora and fauna, the 8th General Assembly of IUCN meeting at Nairobi in 1963 recommends that the practical and political problems involved in illegal export be studied and that an international convention on regulations of export, transit and import of rare or threatened wildlife species or their skins and trophies be drafted and submitted for the approval of governments by the appropriate international organisations possibly on the occasion of a world-wide conference convened for that purpose.

The record of the meeting does not state the proponent of this resolution. Some histories report that Burhenne presented the draft resolution to the General Assembly,[12] but Lee Talbot, who was attending the meeting as a special observer participating in a personal capacity, recalled that he brought the proposal to the meeting where it was presented as a resolution.[13]

In IUCN's programme of work, adopted at the eighth General Assembly, it was agreed that the Union's scientific Survival Service Commission (now Species Survival Commission) would organize a joint symposium with the International Union of Directors of Zoological Gardens on control of export and import of rare or endangered species. This workshop was held in June 1964. It formally adopted two conclusions, one of which had the following recommendation:

effective governmental control of the importation and transit of rare animals. An essential part of such control should be to set up in each country an expert committee to advise Governments as to the species to which this control should be applied.[14]

This text again focussed on the importation end of the trade, with its reference to an 'expert committee' echoing the UK's Animals (Restriction of Importation) Act which was to come into effect a few weeks later.

Within IUCN, work on the development of a draft international convention was being led by its Legislation and Administration Commission, but progress was slow. The Commission lacked funding and staff to advance the work. However, a very first draft outline of a Convention on the Import, Export and Transit of Rare and Vanishing Animals and Plants was prepared in 1964.[15] At IUCN ninth General Assembly in June–July 1966, it was reported that this draft had been prepared with a view to its detailed development and possible consideration at a 'World Conference on Conservation' being organized by United Nations Educational, Scientific and Cultural Organization (UNESCO) in 1967/68.[16] However, the recommendations of that UNESCO conference do not mention trade in wildlife.[17]

A further IUCN draft, titled as a Convention on the Import, Export and Transit of Certain Species was presented to the IUCN Executive Board in April 1967. In September 1967, after internal discussions, a formal version, ready for distribution outside IUCN, was sent via diplomatic channels to 90 countries (with representations in Switzerland, the headquarters) and to various international agencies and organizations.[18] This draft included lists of species whose international trade needed to be better regulated, prepared by the IUCN Survival Service Commission using information from the Red Data Book (now the IUCN Red List of Threatened Species), established in 1964. As reported to the tenth IUCN General Assembly in 1969 the draft received comments from 39 governments, 18 organizations, and various IUCN member organizations which were incorporated into a revised text. Circulation of the revised text was made in August 1969 to the authorities of the 90 countries and to international agencies and organizations, with a request for them to transmit the document to their governments and let IUCN know their position on it.[19] The draft Convention aimed at controlling the unlawful movement of listed species through export and import permits and its key features were summarized:

1. No species (specimens and products) of the listed species may be exported without a permit.
2. Only on presentation of the permit will importation be allowed.
3. At the request of the exporting country these measures may apply to non-listed species in specific cases.
4. An advisory committee be set up to:
   • help in identifying species;
   • perform specific studies on the basis of the export permits delivered;
   • intervene when these studies show that the export rate from any one country in certain species is too high.
5. Sanctions; to stop exports from countries which do not conform to the Convention are provided.

This draft text of a Convention text needed an intergovernmental meeting before it could become a reality. Developments in the USA provided the spark needed for such a meeting. In 1969, the US Endangered Species Preservation Act of 1966 was amended to become the Endangered Species Conservation Act with its provisions expanded to prohibit the commercial import of wildlife threatened with worldwide extinction. In 1967, IUCN was called upon to provide expert testimony to a hearing about the proposed amendments, reflecting their wildlife conservation expertise and their previous consideration of an international convention. The invitation was also the result of a nexus of personal connections between USA officials, IUCN, and Wolfgang Burhenne. In the USA, Stewart Udall, Secretary of the Interior; Russell Train, Under Secretary at the Department of the Interior; and Lee Talbot, by then at the Smithsonian Institution, were key contacts. Burhenne testified, conveying his belief about the need for international cooperation and attaching a copy of the draft Convention which IUCN had prepared.[20]

Commercial interests in the USA complained that the new Endangered Species Conservation Act put them at a competitive disadvantage,[21] and when the law was codified, its Section 5 contained two paragraphs paving the way for the development of an international convention:

(b) To assure the worldwide conservation of endangered species and to prevent competitive harm to affected United States industries, the Secretary, through the Secretary of State, shall seek the convening of an international ministerial meeting on fish and wildlife prior to June 30, 1971, and included in the business of that meeting shall be the signing of a binding international convention on the conservation of endangered species.

(c) There are authorized to be appropriated such sums, not to exceed $200,000, as may be necessary to carry out the provisions of subsection (b) of this section, such sums to remain available until expended.[22]

IUCN's renamed Commission on Legislation continued to work on the text of the draft convention. This was undertaken by Françoise Burhenne-Guilmin with husband Wolfgang Burhenne at the IUCN Environmental Law Centre in Bonn, Germany, and IUCN's Deputy Director General Frank Nicholls at IUCN headquarters in Morges, Switzerland.[23]

Momentum increased with the circulation of the third formal draft in March 1971, which had been discussed with the Food and Agriculture Organization of the United Nations (FAO), the Customs Cooperation Council, the Scientific Committee on Antarctic Research, and the Secretariat of the General Agreement on Tariffs and Trade. The United States indicated its willingness to host the international conference foreseen in its Endangered Species Conservation Act of 1969,[24] and later in the year the Swiss Government offered to act as the depositary for any new convention agreed.[25]

All was not plain sailing though. A number of African states were concerned about the direction that the discussions were taking. They were keen to retain the harvesting quota systems they had been using and were concerned that the species to be protected were to be decided by an 'expert committee'. They produced a Kenya Draft.[26] This followed the despatch of an expert from the US National Audubon Society to Kenya to assist with producing a draft which was more sensitive to developing countries and to conservation. The United States themselves were reported to find the IUCN draft at the time 'a very European document, influenced by European industry and Customs'.[27]

Whilst work continued on the development of a final negotiating text, the United States' offer to host an intergovernmental meeting was delayed. One of the hurdles was said to be the ongoing debate in the United Nations (UN) about which entity should represent China in the UN.[28] In October 1971 when the People's Republic of China was recognized as the only lawful representatives of China by the UN General Assembly, the scene seemed to be set. In the event, neither the Peoples Republic of China nor the Republic of China were listed as attending the plenipotentiary conference, although a representative from the Republic of China is recorded as one of the signatories in the final act.[29]

Further focus was given at the June 1972 UN Conference on the Human Environment held in Stockholm, Sweden, which was the UN's first major conference on the issue of the environment. In its Recommendation 99.3, the Conference called for

> a plenipotentiary conference be convened as soon as possible, under appropriate governmental or intergovernmental auspices, to prepare and adopt a convention on export, import and transit of certain species of wild animals and plants.

In response to the Kenya Alternate Draft of the IUCN text and its own reservations as a significant exporter of wildlife such as alligators (*Alligator* spp.) and bobcat (*Lynx rufus*) skins, the United States, in July 1972, produced a document that compared the IUCN and Kenyan documents and proposed a compromise text.[30]

The United States government convened a *Plenipotentiary Conference to Conclude an International Convention on Trade in Certain Species of Wildlife* to be held at the Pentagon in Washington, DC, from 12 February to 2 March 1973 (Figure 1.1).

All 132 Member States of the United Nations at the time were invited to attend and representatives from 80 did so. These were drawn from all parts of the world, but with an especially strong representation from Africa and Europe. A further eight countries attended as observers. Also represented as observers were five intergovernmental organizations (Customs Cooperation Council, European Communities, FAO, IUCN and the UNESCO) and a single non-governmental organization; the International Council for Bird Preservation. The Conference was chaired by Christian Herter Jr., at the time deputy assistant secretary of state for environmental and population affairs in the US State Department. IUCN continued to play a strong role, providing four of the secretarial staff for the meeting. The chair of the main Drafting Committee, Duncan Poore, was a British scientist who had served as a member of the IUCN Executive Committee from the late 1960s and went on to become its scientific director and then acting

FIGURE 1.1 A view of the Plenipotentiary Conference.

Photo: CITES Secretariat

director-general. The main Drafting Committee was supported by principal committees addressing the lists of animals and of plants to be listed in the Convention's Appendices and a committee on customs matters.[31]

Twenty-three plenary sessions were held over the three weeks of the Conference. The broad structure of the Convention had been assembled during the consultations over the drafts circulated by IUCN in the previous ten years. Nonetheless, extensive discussion of some of the detail was required. The composition of the lists of species whose trade was to be regulated remained an active point of discussion, as it has been ever since. Amongst the other more difficult issues resolved were the following: which types of parts and derivatives of wildlife should be covered by the Convention, the treatment of personal possessions being moved across international borders, the identification and marking of individual specimens in trade, and the way that the list of species covered by the Convention could be amended.[32] Almost all of these technical matters remain difficult problems for the implementation of the Convention 50 years later.

The new international agreement had not been negotiated under the auspices of the United Nations, but as provided for in the Vienna Convention,[33] it became a free-standing treaty governed by international law.

Its preamble recognizes the intergenerational responsibility to protect the world's fauna and flora, which are valuable for reasons varying from aesthetic to economic. The inclusion of both these reasons is significant, as together they encompass both the homocentric approach to nature, which sees species as a commodity, and the intrinsic value approach which maintains that species have an inherent value which must be respected. These different approaches are still reflected in debates and discussions in CITES forums today. The preamble also contains differing reflections about where responsibility lies when it comes to the conservation of species. It maintains, on the one hand, that peoples and States are and should be the best protectors of their own wild fauna and flora, but acknowledging also that international co-operation is essential when it comes to regulating cross-border trade. Careful choice of words in the preamble means that every strand of opinion can find an echo of their own views. The preambular text also provides an overarching purpose against which the efforts of its signatory Parties can be judged:

> the protection of certain species of wild fauna and flora against over-exploitation through international trade

Like some of its forerunners, the Convention divides the species that it addresses into different categories depending on the perceived degree to which they are threatened with biological extinction from international trade.

Appendix I includes species currently threatened with extinction.

Appendix II includes species not necessarily threatened with extinction, but in which trade must be controlled in order to avoid utilization incompatible with their survival.

Appendix III contains species, where one or more Parties responsible for their protection, in their own jurisdiction, considers trade needs to be better regulated to help them achieve their goals.

The species included in Appendices I and II may be amended from time to time by the Parties, and any Party can enter a formal reservation in relation to a specific species listing, either when joining the Convention, or when a change is made to the Appendices. If a Party lodges a reservation, it has no obligation to implement the provisions of the Convention with regard to international trade in that species.

International trade in specimens of Appendix I species is permitted only in exceptional circumstances. International trade in specimens of species included in Appendix II is permitted if the specimen involved was legally obtained and if the export will not be detrimental to the survival of the species. Finally trade in specimens of species included on Appendix III must only take place if the specimen was legally obtained in the territory of the Party requesting its control, or upon proof that the specimen did not originate from the concerned Party.

The types of specimens from these species whose international trade is regulated varies slightly depending on which Appendix the species concerned are included, and whether they are animals or plants.

The fundamental trade controls required under the Convention are a system of permits and certificates. These are to be presented to border authorities upon the import, export, or introduction from the sea (if taken in waters not under the jurisdiction of any State), of specimens of species listed in the Appendices.

However, underneath this simplicity lies a web of special procedures. Exceptions to exempt, or partially exempt, certain specimens of species are applied to specimens in transit; those acquired before CITES provisions applied to them; personal or household effects; animal specimens bred in captivity or plant specimens artificially propagated; specimens destined for scientific research; or those forming part of a travelling collection or exhibition. All are subject to separate protocols of varying complexity.

The basic permits and certificates are issued by each Party's national CITES 'Management Authority', upon advice, where required, from their separate CITES national 'Scientific Authority'.

Live specimens must be transported in a way that minimizes the risk of injury, damage to health, or cruel treatment.

Parties are required to prohibit international trade in specimens of species listed in the Appendices, if not in compliance with the Convention, and both penalize offenders and confiscate such specimens. If the specimens are live, they should be placed in a rescue centre or returned to the State of export.

Parties should report annually on the trade that they have allowed, and submit a biennial report on the legislative, regulatory, and administrative measures that they have taken to implement the Convention.

The Convention specifically provides for Parties to take measures which are stricter than those provided for in the Convention.

Crucially, the Convention text provides for two permanent administrative structures to manage the Convention as a whole on behalf of its Parties. Firstly, a prime governing body—a Conference of the Parties (CoP)—to review progress on the restoration and conservation of the species covered by the Convention and make recommendations for improving its effectiveness and adjusting the lists of species, the trade in specimens of which it regulates. The CoP also makes provisions for the other structure, the Convention's Secretariat, to be provided by the Executive Director of the United Nations Environment Programme (UNEP).

The Secretariat serves the Parties and is required to organize their meetings, promulgate the list of species listed in the Convention's Appendices, and provide tools to help identify them. It is also called upon to make recommendations about matters relating to the aims of the Convention, including from the results of studies it might undertake for this purpose. The Secretariat may also propose remedial action if it finds the Convention is not being properly implemented by a Party.

The Convention requires the Conference of the Parties to meet regularly and the Secretariat to be established permanently in place. Between them, these two bodies have ensured that the dry text of the Convention is brought to life, Parties implement it, and the work of the Convention is made known to a wider audience. The way that these structures have developed and functioned has been critical to the success and longevity of the Convention.

The text of the Convention confirms acceptation of Switzerland's offer to become the Depositary Government for the Convention and consequently to undertake the tasks specified in Article 77 of the Vienna Convention on the Law of Treaties.[34]

The original text of the Convention was established in Chinese, English, French, Russian, and Spanish—each equally authentic. There seemed to be an expectation at the start that the Secretariat should be able to operate in all five of these languages,[35] but at the first meeting of the Conference of the Parties (CoP1) in 1976 the Rules of Procedure designated English, France, and Spanish as the working languages of the meeting. Since then, these three languages have been considered the working languages of CITES.

Further articles of the adopted text deal with joining or leaving the Convention, resolving disputes and amending the Convention.

On Saturday 3 March 1973, the day after the close of the Plenipotentiary Conference, 21 countries signed the Convention on International Trade in Endangered Species of Wild Fauna and Flora, also known as the Washington Convention, particularly in francophone countries. However, the Conference had decided that the Convention would come into force 90 days after the date of deposit by a State of the tenth instrument of ratification. Fittingly perhaps, on 14 January 1974, the USA became the first country to ratify the Convention[36] followed by Nigeria, Switzerland, Tunisia, Sweden, Cyprus, the United Arab Emirates, Ecuador, and Chile. When Uruguay deposited its instrument of ratification on 2 April 1975 the Convention could come into force on 1 July 1975.

Shortly after the Plenipotentiary Conference concluded, the United States Department of State produced a pamphlet titling the meeting as a 'World Wildlife Conference',[37] a name that was to be resurrected many years later by the CITES Secretariat.

The Convention text agreed at the Plenipotentiary Conference provided a solid legal framework for controlling international trade in specimens of the species listed in its Appendices, but it became clear that some further refinements were needed.

1.2 AMENDMENTS TO THE CONVENTION

During the first decade of the Convention's implementation, using the provisions of its Article XVII, two amendments were agreed to Convention text.

The first amendment arose following changes in financial support for the Convention decided by the 6th session of the Governing Council of UNEP on 24 May 1978. The Governing Council decided that its open-ended funding of the operation of CITES should be gradually reduced and stop as soon as possible and no later than the end of 1983.[38] In response, more than one-third of CITES Parties called for an extraordinary meeting of the Conference of the Parties, which was held in Bonn, Germany, on 22 June 1979, in the margins of the Conference to Conclude a Convention on the Conservation of Migratory Species of Wild Animals, where a response of Parties was agreed. At that meeting the Conference of the Parties adopted an amendment to the text of the Convention inserting the words 'and adopt financial provisions' at the end of Article XI, paragraph 3. a), such that the CoP could establish financial arrangements, in particular to support the work of the CITES Secretariat. This amendment entered into force on 13 April 1987, 60 days after two-thirds of the Parties who were members at the time of the Bonn meeting had deposited their formal instruments of acceptance of the amendment with the Depositary Government.

The second amendment to the text of the Convention was agreed on 30 April 1983 at an extraordinary meeting held at the end of CoP4. It amended Article XXI to allow the accession to the Convention by regional economic integration organizations—such as the European Union (European Communities at the time). It did however take over 30 years for this amendment to come into effect, when a sufficient number of Parties who were members in 1983 had deposited their formal instruments of acceptance. It finally came into effect on 29 November 2013.

Several other relatively minor amendments to the text of the Convention have been identified as needing to be put on the agenda of the next extraordinary meeting of the Conference of the Parties, whenever this may be convened. Most were flagged up at CoP1 in 1976,[39] but they have still not been addressed, notwithstanding the two extraordinary meetings held since then:

a) the provisions of Article XVI, regarding the listing of Appendix-III parts and derivatives, should be brought into line with Convention procedures for Appendices I and II (Article XV);

b) paragraph 5 of Article XIV should read: 'Notwithstanding the provisions of Article IV, any export of a specimen' etc.;

c) paragraphs 3 (b) and 5 (b) of Article III should include 'either a Management Authority or a Scientific Authority of the State' etc.;

d) the adoption of an official text of the Convention in Arabic; and

e) correction of errors of an orthographical nature discovered in the text of the Convention.[40]

Although extraordinary meetings can be held back-to-back with regular meetings of the CoP, to be convened, they do require a written request from at least one-third of the Parties. None of the issues noted for possible amendment are of particular concern for specific Parties, hence perhaps the inaction. The adoption of an official text of the Convention in Arabic (which was only added in 2013 as an issue requiring an amendment to the text) may in time prove to be an exception.

NOTES

1 IPBES (2022) *The Thematic Assessment Report on the Sustainable Use of Wild Species of the Intergovernmental Science-Policy Platform on Biodiversity and Ecosystem Services.* Fromentin, J.M., Emery, M.R., Donaldson, J., Danner, M.C., Hallosserie, A., Kieling, D., Balachander, G., Barron, E.S., Chaudhary, R.P., Gasalla, M., Halmy, M., Hicks, C., Park, M.S., Parlee, B., Rice, J., Ticktin, T. and Tittensor, D. (Eds.). IPBES Secretariat, Bonn, Germany. https://doi.org/10.5281/zenodo.6425599.

2 Gilboa, A. and Namdar, D. (2015) On the Beginnings of South Asian Spice Trade with the Mediterranean Region: A Review. *Radiocarbon,* 57 (2): 265–283. https://doi.org/10.2458/azu_rc.57.18562. S2CID 55719842 (accessed 26.02.25).

3 van Uhm, D.P. (2016) *The Illegal Wildlife Trade: Inside the World of Poachers, Smugglers and Traders.* Springer International Publishing AG.

4 Hinsley, A., Willis, J., Dent, A.R., Oyanedel, R., Kubo, T. and Challender, D.W.S. (2023) *Trading Species to Extinction: Evidence of Extinction Linked to the Wildlife Trade.* Cambridge Prisms: Extinction, Vol. 1, e10. https://doi.org/10.1017/ext.2023.7.

5 Boase, T. (2018) *Mrs Pankhurst's Purple Feather: Fashion, Fury and Feminism— Women's Fight for Change.* Arum Press; Folkestone, UK.

6 Bean, M.J. (1983) *The Evolution of National Wildlife Law.* Second edition. Praeger.

7 International Union for the Protection of Nature (1952) *Proceedings and Reports of the third General Assembly, Caracas, Venezuela, 3 to 9 September 1952.* International Union for the Protection of Nature, Brussels, Belgium.

8 International Union for the Conservation of Nature and Natural Resources (1960) *Seventh General Assembly, Warsaw, June 15–24 June 1960.* Proceedings. IUCN. Brussels.

9 Anon (undated) Lee M. Talbot, Ph.D., Endangered Species Coalition. www.endangered.org/campaigns/wild-success-endangered-species-act-at-40/lee-m-talbot/ (accessed 26.02.25).

10 Holgate, M. (1999) *The Green Web. A Union for World Conservation.* Earthscan, London.

11 International Union for the Conservation of Nature and Natural Resources (1964) *Eighth General Assembly Nairobi, Kenya, September 16–24 September 1963.* Proceedings. IUCN Publications New Series. Supplementary papers N°1. IUCN, Morges, Switzerland.

12 Lausche, B.J. (2008) *Weaving a Web of Environmental Law: Contributions of the IUCN Environmental Law Programme.* IUCN/ICEL, Bonn, Germany.

13 Cassel, C. (2013) *Untitled. Fish & Wildlife News Winter 2013.* https://webarchive.library.unt.edu/web/20161230231844/www.fws.gov/international/cites/cop16/fws-news-interview-with-lee-talbot.pdf (accessed 26.02.25).

14 International Union for the Conservation of Nature and Natural Resources (1964) *Symposium "Zoos and Conservation" Report.* IUCN Publications, New Series: Supplementary Paper No. 3. London.

15 Boardman, R. (1981) *International Organization and the Conservation of Nature.* Macmillan Press, London and Basingstoke.

16 International Union for the Conservation of Nature and Natural Resources (1967) *Ninth General Assembly, Lucerne, Switzerland 25 June–2 July 1966.* Proceedings. IUCN Publications New Series. Supplementary Papers N°8.

17 United Nations Educational, Scientific and Cultural Organization (1970) *Use and Conservation of the Biosphere: Proceedings of the Intergovernmental Conference of Experts on the Scientific Basis for Rational Use and Conservation of the Resources of the Biosphere, Paris, 4–13 September 1968*. UNESCO, Paris, France. https://files.eric. ed.gov/fulltext/ED047952.pdf (accessed 26.02.25).

18 Lausche, B.J. (2008) *Weaving a Web of Environmental Law: Contributions of the IUCN Environmental Law Programme*. IUCN/ICEL, Bonn, Germany.

19 International Union for Conservation of Nature and Natural Resources (1970) *Tenth General Assembly Volume II—Proceedings and Summary of Business*. IUCN Publications New Series Supplementary Paper No 27.

20 Adam, R. (2014) *Elephant Treaties: The Colonial Legacy of the Biodiversity Crisis*. University Press of New England, Hanover, USA.

21 Sand, P.H. (1997) Whither CITES? The Evolution of a Treaty Regime in the Borderland of Trade and Environment. *European Journal of International Law*, 8 (1): 29–58.

22 *Endangered Species Conservation Act*. Public Law 91–135—Dec. 5, 1969. www.govinfo. gov/content/pkg/STATUTE-83/pdf/STATUTE-83-Pg275.pdf#page=9 (accessed 26.02.25).

23 Lausche, B.J. (2008) *Weaving a Web of Environmental Law: Contributions of the IUCN Environmental Law Programme*. IUCN/ICEL, Bonn, Germany.

24 Boardman, R. (1981) *International Organization and the Conservation of Nature*. Macmillan Press, London and Basingstoke.

25 Lausche, B.J. (2008) *Weaving a Web of Environmental Law: Contributions of the IUCN Environmental Law Programme*. IUCN/ICEL, Bonn, Germany.

26 Lausche, B.J. (2008) *Weaving a Web of Environmental Law: Contributions of the IUCN Environmental Law Programme*. IUCN/ICEL, Bonn, Germany.

27 Holgate, M. (1999) *The Green Web. A Union for World Conservation*. Earthscan, London.

28 Klimke, V. (2015) *A Sustainable Life. Wolfgang E. Burhenne and the Development of Environmental Law*. Privately Published.

29 Anon (1976) *No. 14537 Convention on International Trade in Endangered Species of Wild Fauna and Flora (with Appendices and Final Act of 2 March 1973). Opened for Signature at Washington on 3 March 1973*. United Nations-Treaty Series Vol. 993: 243–417. https:// treaties.un.org/doc/publication/unts/volume%20993/volume-993-i-14537-english.pdf (accessed 26.02.25).

30 Adam, R. (2014) *Elephant Treaties: The Colonial Legacy of the Biodiversity Crisis*. University Press of New England, Hanover, USA.

31 Anon (1976) *No. 14537 Convention on International Trade in Endangered Species of Wild Fauna and Flora (with Appendices and Final Act of 2 March 1973). Opened for Signature at Washington on 3 March 1973*. United Nations -Treaty Series Vol. 993: 243–417. https://treaties.un.org/doc/publication/unts/volume%20993/volume-993-i-14537-english.pdf (accessed 26.02.25).

32 Mitchell, H. (1977) *History of CITES (The Washington Convention on International Trade in Endangered Species of Wild Fauna and Flora)*. Unpublished.

33 Vienna Convention on the Law of Treaties. Article 2 (a).

34 Anon (2005) *Vienna Convention on the Law of Treaties*. United Nations, Treaty Series, Vol. 1155: 331 1969. https://legal.un.org/ilc/texts/instruments/english/conventions/1_1_1969.pdf (accessed 26.02.25).

35 Doc. 1.41.

36 Anon (1974) *Notification to the Signatory-States of the Convention on International Trade in Endangered Species of Wild Fauna and Flora Signed at Washington on March 3. 1973*. Swiss Confederation. www.eda.admin.ch/content/dam/eda/fr/documents/aussenpoli-tik/voelkerrecht/citesnotifications/740131-CITES_e.pdf (accessed 26.02.25).

37 Anon (1973) *World Wildlife Conference: Efforts to Save Endangered Species*. Department of State Publication 8729, General Foreign Policy Series, 279. Washington, DC.
38 Anon (1979) *United Nations Environment Programme Compendium of Legislative Authority*. Vol. 1, Supplement 1, 1978. https://wedocs.unep.org/bitstream/handle/20.500. 11822/30769/LegEnV1sp.pdf?sequence=1&isAllowed=y (accessed 26.02.25).
39 Resolution Conf. 1.5.
40 Resolution Conf. 4.6 (Rev. CoP19).

2 CITES Structure and Governance

2.1 THE PARTIES

States which agree to be bound by the provisions of the Convention are known as Parties. After the Convention came into force on 1 July 1975, many States quickly joined the new Convention, notably from all parts of the world, with 85 States becoming Parties in the first ten years. The reception of diplomatic *notes verbales* conveying the decision of a State to accede to the Convention is handled by the Depositary Government of Switzerland.[1]

The Convention is not self-executing. Each Party must transpose its provisions into its national law in order for them to take effect. In the early years, many Parties struggled to adopt adequate national laws to implement the Convention. Developments over national laws to implement the Convention are addressed later in this book.

Day-to-day implementation of the Convention with Parties is undertaken by national Management Authorities and Scientific Authorities whose establishment is required under Article IX. Some of their duties are detailed in other Articles of the Convention, but many more obligations have been attributed to them in resolutions of the CoP. The full list of duties of these authorities have then been consolidated in resolutions of the CoP, one each for Management[2] and Scientific Authorities.[3] Each Party is free to establish as many Management and Scientific Authorities as it thinks fit. Some have designated many, sometimes in keeping with their federal government structure or national competencies, for example, terrestrial or marine species. However, only one Management Authority is authorized to communicate with other Parties and with the Secretariat and may be termed the lead authority. Initially, wildlife or national park ministries were often designated as the national Management Authorities. In time, these were in some cases replaced by ministries with broader environment or forestry responsibilities. However, some Parties decided to designate trade ministries or those with competence for veterinary affairs as Management Authorities. Likewise, in the early years, natural history museums, botanical gardens, and the like were the main institutions designated as Scientific Authorities. Although not conducive to the proper functioning of the Convention, sometimes the Management Authority was also designated as the Scientific Authority. Occasionally smaller countries who are Parties establish expert committees, serviced by an appropriate ministry, to serve as their Scientific Authority. With the increasing recognition of the importance of wildlife generally and consequent growth in expertise, national biodiversity agencies are now often used as a Scientific Authority. Whilst the identity of the authorities has changed, it is hard to escape the conclusion that the resources at their disposal to undertake their duties have remained inadequate. Digital tools and

DOI: 10.1201/9781003542278-2

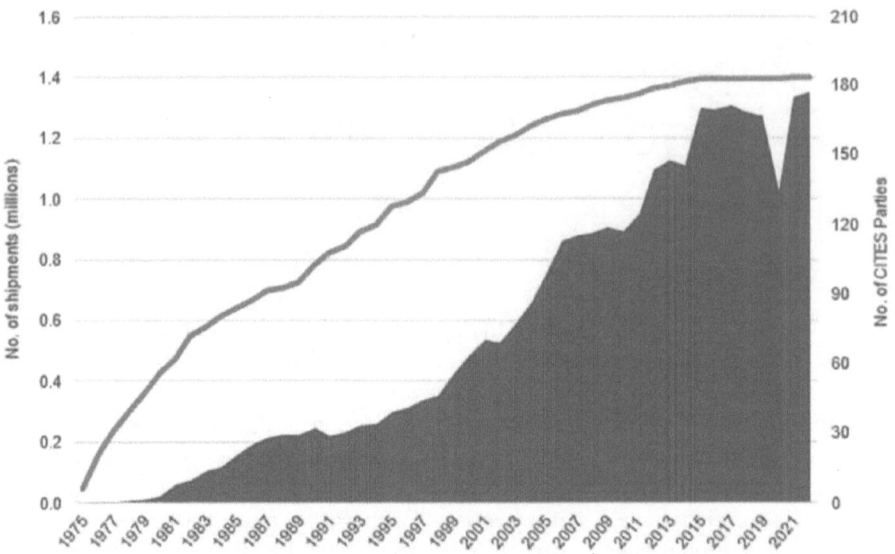

FIGURE 2.1 The number of trade transactions reported per year, 1975–2022 (in millions) and the number of Parties to the Convention (cumulative line) since CITES entered into force in 1975.

Source: UN Environment Programme World Conservation Monitoring Centre. CITES Trade Database, data extracted 28 January 2025

computerized permitting have helped, but the number of CITES permits and certificates that Parties issue has grown relentlessly. Well over a million records of trade in CITES-listed wildlife species are now reported by Parties annually, and the database that records them contains more than 27 million records (Figure 2.1).

The reduced financial resources for Management Authorities reflects declining resources for public administrations around the world generally. National representatives attending CITES meetings appear to be more junior, as senior staff have increasing portfolios and cannot afford the time to devote to the detail of CITES implementation. One effect of this has also been the outsourcing of some CITES activities, directly or indirectly, often to civil society organizations. National CITES authorities who lead implementation on the Convention are now often constrained to leave development of CITES policy initiatives to these organizations and are content to present the results to the other Parties.

A significant obligation for national CITES authorities is to report on actions taken to implement the Convention. The text of the Convention requires two sorts of reporting: annual reports containing details of the trade that it has authorized under the Convention and a biennial report on legislative, regulatory, and administrative measures taken to enforce the Convention.[4] However, subsequent resolutions and decisions of the CoP have added dozens (a total of over 80 after CoP19 in 2022) of other reporting requirements, mostly relating to actions taken to implement the Convention.[5] Some are very specific, apply to only a selection of Parties, and may be periodic; others, such as an annual illegal trade report, apply to all Parties and represent a significant undertaking. Repeated attempts to reduce the reporting burden on Parties have met with very limited success. As in other fields related to the implementation of the Convention, all proposed new activities have their strong supporting Parties, and other Parties hesitate to oppose extra reporting obligations for fear of appearing too negative light or being wary that their own special interests may also fall victim to rationalization later. This difficulty in prioritization and inability to match actions against the resources available to deliver them has increasingly bedevilled the Convention. Annual trade reports are essential for monitoring the volume of trade being authorized for individual species. This is analyzed for signs of unsustainable trade under a Review of Significant Trade process which is described in more detail later in this book. In the early years of the Convention, Parties had the idea of producing a regular comparative tabulation of annual report data in the form of a Yearbook of International Wildlife Trade,[6] but no progress was made due to lack of resources. After being dropped at CoP9 in 1994, the idea was resurrected at CoP12 in 2002,[7] but it was not until 2022 that a pilot World Wildlife Trade Report was produced,[8] the fruits of a broad partnership between the Secretariat and UNEP, the United Nations Conference on Trade and Development, the World Trade Organization, IUCN, and TRAFFIC. Arguably the idea of a yearbook has been overtaken by online tools, such as the CITES Wildlife TradeView,[9] which allows a bespoke analysis of any aspect of international trade regulated by the Convention.

The main duties of the Scientific Authorities are to advise the Management Authority on whether or not an export will be detrimental to the survival of that species and, in the case of any import of a species included in Appendix I, whether or not that import will be for purposes which are not detrimental to the survival of the species involved. For exports of specimens of Appendix II species, the Scientific Authority must monitor the export permits granted and the actual resulting exports of such specimens. They should then advise on suitable measures to be taken to limit the grant of export permits if the levels of trade would not maintain the species throughout its range at a level consistent with its role in its ecosystem. For imports of living specimens of Appendix I species, the Scientific Authority should also satisfy itself that the proposed recipient is suitably equipped to house and care for them.

Although not mentioned in the text of the Convention, Parties are also encouraged by resolution to establish enforcement focal points with expertise in online investigations, evidence gathering and prosecutions, to correspond with other Parties and intergovernmental organizations on this matter.[10]

One Party that stands apart from the others because of its nature is the European Union (EU). Representatives from the European Communities (a forerunner of the EU) were present at the plenipotentiary Conference to conclude CITES in 1973, but the Convention was not drafted with what is termed regional economic integration organizations (REIOs) in mind. One of the European Communities, the European Economic Community (EEC), wasted no time in addressing this issue, commissioning IUCN in 1976—in collaboration with the Secretariat—to prepare a membership feasibility study.[11]

At CoP1 in 1976, European Commission official Claus Stuffmann, who went on to play a major role in getting the EU behind CITES, noted that the EEC should be regarded as one entity from the view-point of trade, whether or not individual EEC member nations had already ratified the Convention, and expressed the hope that the EEC would become a Party to the Convention.[12] By CoP3 in 1981, some Parties were intervening at the CoP 'on behalf of the EEC'. An extraordinary meeting of the CoP was held at the end of CoP4 in 1983 to consider a proposal to amend the convention to allow accession to the Convention by REIOs where the member states of these organizations have ceded competence over some issues, such as trade negotiations, to such an organization. After clarifying that the number of votes that could be cast by a REIO at a CoP was equal to the number of its member states which are Party to the Convention, the extraordinary meeting agreed the change to the Convention text after a secret ballot, with 27 votes in favour, 9 votes against, and 6 abstentions.[13]

The EEC was moving to abolish all controls on trade on the internal borders between its member states. It passed EEC-wide laws to implement the Convention applying directly to its member states.[14] One immediate impact of this was that Belgium, Luxembourg, and the Netherlands, which were not Party to CITES, were required to join the Convention in 1983/1984, and the same applied to Spain, which joined the EEC and CITES in 1986.

The logistical challenges of applying the Convention to a REIO were considerable. Concern arose amongst other Parties about the impact of EEC measures on the implementation of the Convention by Parties who were also member states of the EEC, and resolutions were adopted calling for improvements.[15] The EEC commissioned a major study of the issue from IUCN's World Conservation Monitoring Centre and Environmental Law Centre[16] and subsequently strengthened its internal legislation on CITES.[17] The concerns subsided, and at CoP12 in 2002 it was decided to repeal the original resolutions calling for improvements. Throughout this time, the EEC engaged in CITES as a block on matters where it had competence—primarily where any trade issues were involved. Initially the REIO had 15 votes at the CoP, one for each of the member States who were Party to the Convention. This rose to 25 votes when the EU was enlarged by new members in 2004. Further, CITES Parties that were hoping to join the EU also tended to align their voting with them.

This gave the EU considerable influence relative to other Parties, and so there was concern about this ascendancy. As a consequence, after CoP11 in 2002, other regions or sub-regions, beginning with the Association of Southeast Asian Nations (ASEAN),[18] whilst not having the legal capabilities of a REIO, began to coordinate their inputs to CoPs, and the influence of the EU, whilst still strong, was tempered. Doubts over the efficacy of the EEC/EU's implementation of the Convention and concerns over its impact on decision-making meant that it was only over 30 years later in 2013, when a sufficient number of States who were Parties to CITES in 1983 had deposited their formal instruments of acceptance, that the amendment to the Convention allowing REIOs to become Parties came into effect. The EU became a Party in its own right in 2015, and currently it is the only block of States which meets the definition of a REIO under CITES.

2.2 THE CONFERENCE OF THE PARTIES

The Conference of the Parties (CoP) called for in the text of the Convention is the supreme governing body for the Convention and has proven to be a key driver for its development and evolution. Representatives of all Parties who have acceded to or ratified the Convention may attend and participate in decision-making, and in keeping with the open spirit of the Convention, representatives from United Nations bodies, States not Party to the Convention and bodies or agencies 'technically qualified in protection, conservation or management of wild fauna and flora' may attend as observers. In practice, virtually all requests to attend as an observer are accepted, including from trade organizations. Such an open approach to civil society participation in meetings of a multilateral environmental agreement was pioneering at the time of its introduction. CoP3 in 1981 noted the considerable increase in the numbers of observer organizations attending the meetings of the CoP, which resulted in additional expenditure and decided to apply a standard participation charge of USD 50 for all non-UN observer organizations.[19] This fee has risen substantially to USD 600 for the first participant and USD 300 for each additional participant at CoP19 in 2022. These fees add to the considerable cost of sending delegates to a two-week conference, generally involving international travel as well. However, this has not deterred Western conservation and animal welfare non-governmental organizations (NGOs) in particular, who continue to send large numbers of delegates to CoPs. As for Parties, not all find it easy to access the funds to participate at meetings of the CoP, and for that reason the Secretariat initiated a Sponsored Delegates Project (SDP) in 1986 to solicit financial assistance from governments and organizations for a fund that provides 'anonymous' assistance for travel, accommodation, and a daily subsistence allowance to enable developing country representatives to attend meetings of the CoP (Table 2.1).

On average, over half of all Parties attending meetings of the CoP have at least one delegate paid by the SDP since it began, although sometimes the same Parties send a number of additional delegates using other funding of their own, or that provided by others. Parties left the task of deciding which countries should have delegates sponsored to the Secretariat which has in recent years based its selections on the following:

• Human Development Index ranking from highest to lowest
• Least Developed Countries and/or Small Island Development States
• Parties with less than six delegates at the previous meeting of the CoP
• Parties who are members of the Standing Committee
• Parties on the agenda in relation to compliance matters for the Standing Committee meetings held in the margin of the CoP
• Submission of an amendment proposal for the CoP in question

For many years the SDP was only a Secretariat initiative, but in 2016 the Secretariat suggested that it should be formalized through a resolution of the CoP.[20] Notwithstanding the existence of the SDP, the Secretariat noted that funding of delegates directly by other Parties and by other donors, including NGOs and foundations, could give the

TABLE 2.1

Number of Sponsored Delegates and Parties Having a Sponsored Delegate at Meetings of the CoP

CoP	Number of sponsored delegates	Number of Parties having a sponsored delegate
CoP6 (1987)	113	No data
CoP7 (1989)	150	No data
CoP8 (1992)	152	No data
CoP9 (1994)	175	No data
CoP10 (1997)	196	No data
CoP11 (2000)	207	108
CoP12 (2002)	154	100
CoP13 (2004)	149	97
CoP14 (2007)	181	114
CoP15 (2010)	126	94
CoP16 (2013)	162	87
CoP17 (2016)	135	102
CoP18 (2019)	93	94
CoP19 (2022)	101	67
Total/Average per CoP	2094/150	863/96

perception of undue influence and that the Convention needed to be seen to be operating in a fully open and transparent manner. A resulting resolution[21] therefore called on donors to support delegates through the SPD rather than bilaterally and urges both those giving and receiving sponsorship outside the SDP to inform the Secretariat so that this information can be made public. At CoP18 in 2019, 24 delegates were reported as receiving such bilateral sponsorship and 13 such delegates at CoP19 in 2022.[22]

According to Article XI of the Convention, unless the Conference decides otherwise, CoPs are to be held at least once every two years, but this periodicity has only been achieved three times in 19 CoPs, and in recent years organizational difficulties have extended the inter-CoP interval to up to three and a half years (Table 2.2). Article XI.1 required that a first meeting of the CoP should be held not later than two years after the Convention's entry into force (i.e., 30 June 1977) and its eventual timing, from 2–6 November 1976, reflected a compromise between those Parties who wished to try and resolve some of the implementation issues quickly and those who thought more experience was required before this could be done.[23] At CoP1, it was quickly realized that more detailed discussions were needed on a number of technical issues:

- Tools to identify specimens
- Guidelines on the care and shipment of live specimens
- Which parts and derivatives of listed species are readily recognizable
- Exchange preserved animal specimens between registered scientific institutions
- The scientific validity of the species listings in Appendices I and II

Consequently, it was agreed to hold a non-plenary Special Working Session of the Conference of the Parties on implementation issues[24] involving technical experts in the fields of law enforcement, customs control, zoology, botany, and the veterinary sciences. The meeting was held in Geneva, Switzerland, from 17–28 October 1977 and prepared a report[25] which served as the basis for many agreements adopted at CoP2 held in San José, Costa Rica, in 1979.

TABLE 2.2
Venues, Dates, and Participants at Meetings of the Conference of the Parties

	Venue	Year	Number of Parties at the time	Number of Parties participating	Observer organizations
CoP1	Berne, Switzerland	1976	30	24	27
CoP2	San José, Costa Rica	1979	47	33	70
CoP3	New Delhi, India	1981	61	54	97
CoP 4	Gaborone, Botswana	1983	78	59	77
CoP 5	Buenos Aires, Argentina	1985	85	67	128
CoP 6	Ottawa, Canada	1987	92	87	114
CoP 7	Lausanne, Switzerland	1989	101	94	150
CoP 8	Kyoto, Japan	1992	114	104	163
CoP 9	Fort Lauderdale, USA	1994	124	117	153
CoP10	Harare, Zimbabwe	1997	136	129	164
CoP11	Gigiri, Kenya	2000	149	136	180
CoP12	Santiago, Chile	2002	160	142	148
CoP13	Bangkok, Thailand	2004	166	152	136
CoP14	The Hague, the Netherlands	2007	172	154	158
CoP15	Doha, Qatar	2010	175	157	132
CoP16	Bangkok, Thailand	2013	178	164	192
CoP17	Johannesburg, South Africa	2016	183	160	182
CoP18	Geneva, Switzerland	2019	183	172	212
CoP19	Panama City, Panama	2022	184	152	194

Meetings of the CoP were almost invariably 10 working days, with two rest days in the middle. For CoP16 in 2013, to limit costs, it was decided that in future this be reduced to 9 working days,[26] but this was premised on the observation that the total number of agenda items discussed at CoPs had not varied enormously between 2002 and 2010.[27] Since then, the number of agenda items considered at meetings of the CoP has increased by 30–50%, but given budgetary constraints, it seems unlikely that there will be any return to longer meetings or increased frequency.

The basic structure of the meetings of the CoP has remained remarkably stable since the very first meeting, which is testament to the insight of the early representatives of the Parties and the Secretariat. Most of the final decisions made by Plenary sessions are based on recommendations from two sessional committees which are normally held concurrently, but when the plenary is not in session. These sessional committees broadly covered scientific issues on one side and administrative, implementation, budgetary, and legal issues on the other. Until CoP6 in 1987, the scientific committee (known as 'Committee I') was sometimes termed the screening group/committee (usually divided into animals and plants sub-committees), as its main task then was to review proposals to amend the Appendices. A screening sub-committee for ranching proposals (explained later in this book) was also created at CoP4 in 1983. The other committee (known as 'Committee II') was initially formed of the existing Technical Expert Committee (described later in this chapter), but when that Committee was dissolved, by a committee established at each CoP. A Credentials Committee has been established at every meeting of the CoP to examine the credentials presented by delegates and recommend their acceptance if they conform to the standards agreed by the Parties. The formal CoP 'officers' were initially just the Plenary chair and vice-chairs but at CoP6 in 1987 this was expanded to include the Chairs of Committees I and II, and at CoP11 in 2000 to include the Chair of the Credentials Committee (the Chair of the Budget Committee was also listed an officer for CoP11 only). These officers are chosen by the CoP on a recommendation from the Standing Committee after Parties have been invited to suggest possible candidates. The chair of the CoP, who leads the plenary sessions, is by tradition from the formal hosting Party, when there is one (which was not the case for various reasons at CoP7 in 1989, CoP11 in 2000 and CoP18 in 2019). As the chair is often a senior official or minister with other commitments, since CoP14 in 2007, he/she has also had an 'alternate' to replace him/her when he/she was unavailable. The chair of the CoP initially chose the vice-chairs who were drawn from the members of the Steering/Standing Committee (described later in this chapter), but more recently two vice-chairs have been designated by the CoP on a recommendation from the Standing Committee. The only qualification for these CoP officer posts is that candidates are, 'prima facie, capable of impartially expediting the business of the Conference'. As the officers have no vote, there are no other qualifications required in order to be nominated.[28] The list of officers at meetings of the CoP (including those positions which were not considered 'officers' at earlier meetings) shows a preponderance of male, anglophone postholders, and persons from Europe and North America, in particular for the position of chair of Committees I and II.

CoP1–1976

CoP Chair: Mr A. Nabholz (Switzerland)
Vice-Chairs: Mr R. Boden (Australia) and Ms P. Fox (USA)
  *Committee I: Mr V.C. Wynne-Edwards (UK)
  *Committee II: Mr S. B. Shaybani (Islamic Republic of Iran)
  *Credentials Committee: Ms L. Thompson (Canada)

CoP2–1979

CoP Chair: Mr Carlos Villalobos Sole (Costa Rica)
Vice-chairs: Mr J. Heppes (Canada), Mr F. W. King (USA), Mr P Gaffer
  (Switzerland), Mr M. Silberman (Costa Rica), Mr A. Bonilla Ascona (Costa
  Rica), Mr E. Lopez Pizarro (Costa Rica) and Mr E. Ayensu (USA)
  *Fauna Working Group: not named
  *Flora Working Group: not named
  *Credentials Committee: Mr Z. Odero Kongoro (Kenya)

CoP3–1981

CoP Chair: Mr M.R. Dalvi (India)
Vice-Chairs: Mr R. Parsons (USA) and Mr S. Singh (India)
  *Technical Expert Committee: Mr N. Jayal (India)
  *Screening Committee (Animals): Mr P. Gafner (Switzerland)
  *Screening Committee (Botanical Sub-Group): Mr S. Jain (India)
  *Credentials Committee: Ms M. Bhaduri (India)

CoP4–1983

CoP Chair: Mr E. Matenge (Botswana)
Vice-Chairs: Mr S. Singh (India), Mr J. Heppes (Canada) and Mr J. Servat
  (France)
  *Technical Committee: Mr J. Goldsmith (UK)
  *Screening Committee (Animals): Mr W. King (IUCN)
  *Screening Committee (Plants): Ms S. Oldfield (UK)
  *Screening Committee (Ranching): Mr J.D. Ovington (Australia)
  *Credentials Committee: Mr G. A. Furness, Jr. (USA)

CoP5–1985

CoP Chair: Mr E. Gonzalez Ruiz (Argentina)
Vice-Chairs Mr S. Singh (India) and Mr J. Heppes S (Canada)
  *Technical Committee: Mr G. Emonds (Germany)
  *Screening Committee (Animals): Mr G. Child (Zimbabwe)
  *Screening Committee (Plants): Ms S. Oldfield (UK)
  *Credentials Committee: Mr G. A. Furness, Jr. (USA)

CoP6–1987

CoP Chair: Mr D. Munro (Canada)
Vice-Chairs: Mr M.K. Ranjitsính (India) and Mr E. Gonzalez Ruiz (Argentina)
Committee I: Mr P. Dollinger (Switzerland)
Committee II: Mr J.P. Oriero (Kenya)
*Credentials Committee: Mr D. K. Pollock (Canada)

CoP7–1989

CoP Chair: Mr M. Surbiguet (France)
Vice-Chairs: Mr J. Heppes (Canada) and Mr H. Nsanjama (Malawi)
Committee I: Mr P. Dollinger (Switzerland)
Committee II: Mr J. Mendez Arrocha (Venezuela)
*Credentials Committee: Ms Janet Owen (New Zealand)

CoP8–1992

CoP Chair: Mr N. Akao (Japan)
Vice-Chairs: Ms C. James (Trinidad and Tobago) and Mr V. Koester (Denmark)
Committee I: Mr M. Holdgate (UK)
Committee II: Mr M. Jones (USA)
*Credentials Committee: Mr G. J. Fraser (Australia)

CoP9–1994

CoP Chair: Mr F. Loy (USA)
Vice-Chairs: Mr S. C. Dey (India) and Mr G. Doungoubé (Central African Republic)
Committee I: Mr E. Ezcurra (Mexico)
Committee II: Ms V. Lichtschein (Argentina)
*Credentials Committee: Ms S. Wagner (USA)

CoP10–1997

CoP Chair: Mr T. J. Jokonya (Zimbabwe)
Vice-Chairs: Ms N. Nathai-Gyan (Trinidad and Tobago) and Mr H. M.A.
    Tatwany (Saudi Arabia)
Committee I: Mr D. Brackett (Canada)
Committee II: Mr J. Rubio de Urquia (Spain)
*Credentials Committee: Mr S.C. Dey (India)

CoP11–2000

CoP Chair: Mr B. Asadi (Islamic Republic of Iran)
Vice-Chairs: Mr H. Walters (St. Lucia) and Mr E. Severre (United Republic
    of Tanzania)
Committee I: Ms M. Clemente Muñoz (Spain)

Committee II: Mr V. Koester (Denmark)
Credentials Committee: Ms J. Owen (New Zealand)

CoP12–2002

CoP Chair: Mr S. Bitar (Chile)
Vice-Chairs: Uganda and the USA
Committee I: Mr D. Morgan (UK)
Committee II: Ms A-M Delahunt (Australia)
Credentials Committee: Mr M. Lauprasert (Thailand)

CoP13–2004

CoP Chair: Mr S. Khunkitti (Thailand)
Vice-Chairs: Mr D. Brackett (Canada) and Ms V. Lichtschein (Argentina)
Committee I: Ms H. Dublin (USA)
Committee II: Mr M. Brasher (UK)
Credentials Committee: not named (Czech Republic)

CoP14–2007

CoP Chair: Ms Gerda Verburg (Netherlands)
Alternate Chair: Mr A. van de Zande (Netherlands)
Vice-Chairs: Mr M. Jones (USA) and Mr M. Calvar Agrelo (Uruguay)
Committee I: Mr G. Leach (Australia)
Committee II: Mr C-S Cheung (China)
Credentials Committee: Ms S. Meintjes (South Africa)

CoP15–2010

CoP Chair: Mr F. Al-Thani (Qatar)
Alternate Chair: Mr G. Abdullah Mohammed (Qatar)
Vice-Chairs: Ms N. Cespedes (Chile) and Mr R. Gabel (USA)
Committee I: Mr J. Donaldson (South Africa)
Committee II: Mr W. Dovey (New Zealand)
Credentials Committee: Mr G. Evrard (Belgium)

CoP16–2013

CoP Chair: Mr P. Rengsomboonsuk (Thailand)
Alternate Chair: Mr P. Pukkaman (Thailand)
Vice-Chairs: Mr P. C. Wilungula (Democratic Republic of the Congo) and
    Mr Ø. Størkersen (Norway)
Committee I: Ms C. Caceres (Canada)
Committee II: Mr R. Gabel (USA)
Credentials Committee: Ms Z. Zhou (China)

CoP17–2016

CoP Chair: Ms M. Nkoana-Mashabane (South Africa)
Alternate Chair: Ms B.E. Molewa (South Africa)
Vice-Chairs: Mr C. Taolo (Botswana) and Ms. S. Al-Salem (Kuwait)
Committee I: Ms K. Gaynor (Ireland)
Committee II: Mr J. Barzdo (Switzerland)
Credentials Committee: Mr B. Alfaleh (Saudi Arabia)

CoP18–2019

CoP Chair: Mr T. Jemmi (Switzerland)
Alternate Chair: Ms A. Ochieng Pernet (Switzerland)
Vice-Chairs: Mr M. Isaacs (Bahamas) and Mr J. Lutalo (Uganda)
Committee I: Mr R. Hay (New Zealand)
Committee II: Mr C. Hoover (USA)
Credentials Committee: Ms M. van Looy (Belgium)

CoP19–2022

CoP Chair: Mr M. Concepción (Panama)
Alternate Chair: Ms S. Binder (Panama)
Vice-Chairs: Ms P. Gandiwa (Zimbabwe) and Ms. A. Wong (Singapore)
Committee I: Mr V. Fleming (UK)
Committee II: Ms R. Ollerenshaw (Australia)
Credentials Committee: Ms H. Mesbah (Morocco)

* Not formally 'officers' of the CoP at the time of the CoP in question.

Meetings of the CoP are key events for Parties and for their Secretariat (Figure 2.2). Assisted by the SDP, over 80% of Parties regularly attend meetings of the CoP (Figure 2.3).

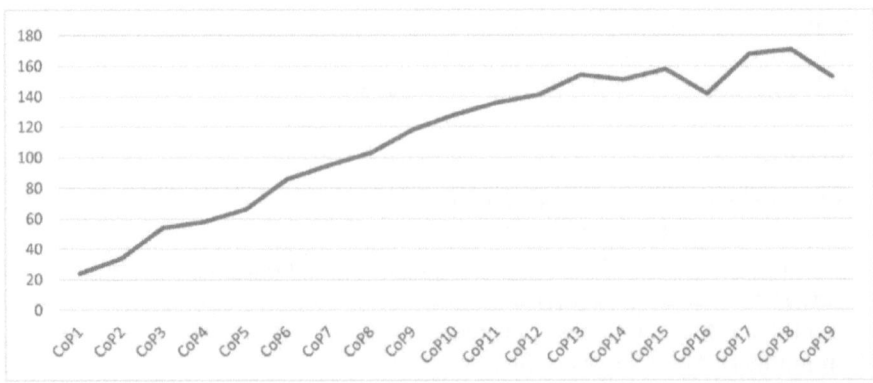

FIGURE 2.2 Number of Parties represented at meetings of the Conference of the Parties.

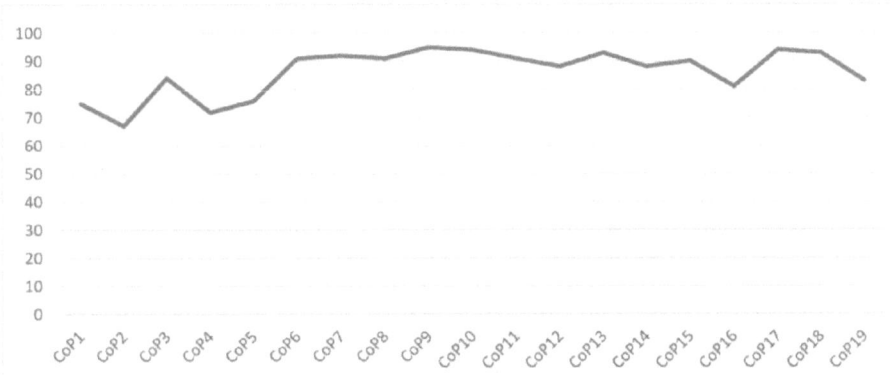

FIGURE 2.3   Percentage of Parties represented at meetings of the Conference of the Parties.

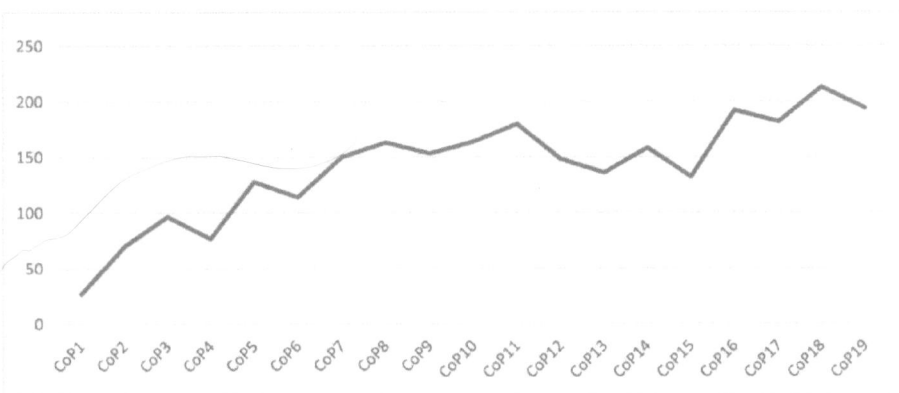

FIGURE 2.4   Number of observer organizations attending meetings of the Conference of the Parties.

The many non-State observer organizations also see meetings of the CoP as an opportunity not only to influence policy, but also to raise funds and profile. Over 1,000 different observer organizations have attended the first 19 meetings of the CoP. Their number has grown significantly, both in terms of number of different organizations and number of delegates (Figure 2.4). In terms of size, the larger ones now have delegations (persons registered to attend the CoP) which comfortably exceed that of most of the Parties.

Some international conventions provide for decisions to be taken by a vote if required, but this option is rarely used. Conventions in the field of biodiversity established more recently, such as the Convention on Biological Diversity, take decisions by consensus, and although this is the objective in CITES, if a consensus on the adoption or rejection of a proposal cannot be found, the presiding officer must put the matter to a vote. This happens more frequently in the case of proposals to amend the Appendices, where the choice to be made by Parties is more binary. At recent meetings of the CoP around one-third of such proposals have been decided after a vote. Other issues decided by the CoP related to the interpretation and implementation of the Convention have often been pre-discussed at meetings of the intersessional committees and come to the CoP with some degree of consensus and do not require a vote. However, skilful chairing also plays a role as demonstrated at CoP18 in 2019, when the business of Committee II was concluded without a single vote. The text of the Convention requires a two-thirds majority of Parties present and voting for amendments to the Appendices and other amendments to the Convention to be adopted. Initially, all other matters were decided at the CoP by a simple majority of the votes cast. In view of the development of the interim financial provisions for the budget of the Convention, the requirement for a two-thirds majority was also extended to the terms of reference for the administration of the Trust Fund at CoP2 in 1979. At CoP5 in 1985, some Parties were concerned that resolutions interpreting the provisions of the Convention could have significant effects—itself a sign of the increasing seriousness with Parties were taking the conclusions agreed by the CoP. So at CoP6 in 1987, on a proposal from the Secretariat, Parties agreed that all votes on procedural matters relating to the forwarding of the business of the meeting should be decided by a simple majority of votes cast, while all other decisions should be by a two-thirds majority of such votes. Voting was normally by a show of hands until electronic voting was introduced at CoP13 in 2004, with a record of the votes cast projected on a big screen. Other types of voting are also permitted under the Rules of Procedure, such as a roll-call vote, where Parties with approved credentials are called upon to indicate their vote one by one or, more controversially, secret ballots. Secret ballots have often been used at meetings of the CoP in relation to proposals to amend the Appendices for controversial or 'political' taxa such as marine species and the African elephant *Loxodonta africana*, or policy decisions related to the ivory trade generally. Advocates say that they protect delegates from undue pressure from non-State observers, detractors that they undermine the spirit of openness and transparency. At the outset, any single Party could request that a vote be held by secret ballot, but at CoP5 in 1985, these were proving cumbersome and time-consuming, and the then Secretary-General announced his intention to propose changes.[29] At the subsequent meeting of the CoP the Rules of Procedure were changed such that all votes in respect of the election of officers or of prospective host countries for CoPs should be held by secret ballot, the need for which was widely understood. For other matters, a secret ballot could be undertaken after being confirmed by a simple majority of votes cast (not itself to be undertaken by secret ballot).[30] The latter provision however was not supported by everyone. At a meeting of the Standing Committee in 1994, Zimbabwe claimed that as an open vote was needed to approve a proposal

FIGURE 2.5   Official CITES logo.

[The use of the CITES logo is for purely illustrative purposes and it does not imply any association of CITES with the contents of this publication. The views expressed herein are those of the author and do not necessarily reflect the views of the CITES Secretariat or the United Nations.]

for a secret ballot, delegations may find themselves under pressure as there would be a suspicion that if they supported a secret ballot, they intended to vote in a certain way on the substantive matter.[31] At the subsequent CoP9 in 1994, the Rules of Procedure were further amended to permit a secret ballot if such a request was seconded by ten Parties. Further attempts to limit the use of secret ballots were made at subsequent CoP meetings,[32] but to date these have been unsuccessful, to the disappointment of many civil society groups.[33]

At CoP3 in 1981 a logo for the meeting was used for the first time. It was the acronym CITES in the form of an elephant and was created by Patrick Virolle, then IUCN's graphic artist, at the request of Switzerland's Peter Dollinger, then the chair of the Identification Manual committee.[34] The logo proved popular, and after a contribution of 2,500 USD by the Association of European Ivory Traders for its design and production, it was officially adopted by the Standing Committee in October 1981[35] (Figure 2.5).

The logo went on to become one of the most distinctive pictorial references for the Convention. Since 1995, the use of the logo has been protected under Article 6t of the Paris Convention for the Protection of Industrial Property and its use is strictly governed by the Secretariat.[36] Parties have been keen to ensure that the logo is used only for non-commercial purposes and that unless utilized for official purposes, it is not used to imply any endorsement by the Convention or the Parties. Host governments for meetings of the CoP started developing their own logos for the meetings. Initially they were quite simple and incorporated the CITES logo used at CoP3, but from CoP12 in 2002 onwards they became more imaginative and an important signature for the meetings (Figure 2.6).

Meetings of the CoP are also an opportunity to demonstrate the rather addictive nature of CITES. Many participants have attended multiple meetings of the CoP, often representing (very) diverse organizations at different meetings. Remarkably, two participants who attended CoP1 in 1976 also participated at CoP19 in 2022 (Table 2.3).

FIGURE 2.6 Logos for meetings of the Conference of the Parties.

[The use of the CITES logo is for purely illustrative purposes and it does not imply any association of CITES with the contents of this publication. The views expressed herein are those of the author and do not necessarily reflect the views of the CITES Secretariat or the United Nations.]

TABLE 2.3

Participants Attending the Greatest Number of Meetings of the Conference of the Parties

Participant	Number of CoPs attended (of 19)	Representing
Eugène Lapointe	18	Canada, the Secretariat, International Wildlife Management Coalition.
Marshall Meyers	18	Laboratory for Experimental Medicine and Surgery in Primates, Pet Industry Joint Advisory Council, Conservation Treaty Support Fund, Ornamental Fish International.
Jaques Berney	17	The Secretariat, International Wildlife Management Coalition.
Bill Clark	17	Friends of Animals, International Donkey Protection Trust, Israel, INTERPOL.
Rick Parsons	17	USA, Pet Industry Joint Advisory Council, Wildlife Coalition International, Safari Club International.
Willem Wijnstekers	16	European Commission, European Economic Community, the Secretariat, CIC—International Council for Game and Wildlife Conservation.
Jonathan Barzdo	15	IUCN, Rapporteur, TRAFFIC, the Secretariat, Switzerland, Jonathan Barzdo Consultant.
Esko Jaakkola	14	Finland.

Source: Proceedings of the meetings of the Conference of the Parties.

2.3   RECORDING THE AGREEMENTS OF THE CONFERENCE OF THE PARTIES

In addition to the changes to the text of the Convention, agreed in its early years at two extraordinary meetings, it quickly became clear that a number of its provisions required clarification for them to be implemented in a harmonious fashion by all Parties. Agreements were needed on the interpretation of various terms used in the text of the Convention and on their application. The natural forums for such agreement to be debated and agreed were the meetings of the CoP. Discussions at the first meeting of the CoP, in Berne, Switzerland, resulted in a number of conclusions. However, these were not all fully or clearly codified at the meeting, and so the Secretariat subsequently turned them into 'Resolutions' of the CoP.[37] The resolutions agreed at the first meeting of the CoP included two omnibus resolutions, which addressed a variety of difficulties in implementation and interpretation of certain provisions of the Convention that had been noted by the Parties in the aftermath of the Plenipotentiary Conference and the first 15 months of implementation.

Resolutions adopted by the CoP are not themselves legally binding on Parties, but there is an expectation that they will be applied by Parties. Peer pressure to implement them is high, and the extension of the Convention's compliance practices to their implementation means that they are taken seriously by a majority of Parties.

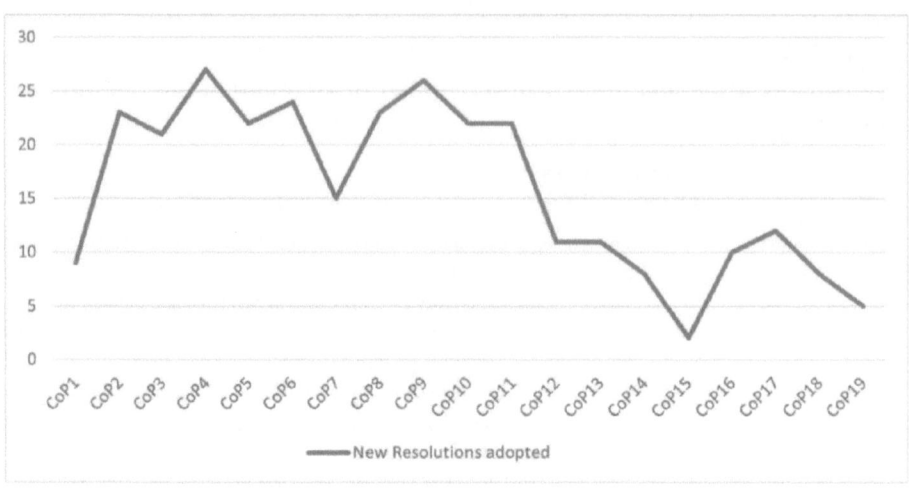

FIGURE 2.7   Number of new Resolutions adopted by each meeting of the Conference of the Parties.

Resolutions of the Conference of the Parties generally refer to long-term agreements of principle concerning the interpretation of certain aspects of the Convention, policies regarding particular species groups, reactions to external developments, or relations with other organizations. Soon after the Convention was established, when much remained to be clarified, up to 27 resolutions were agreed at a single CoP, but from CoP11 in 2000 their number declined (Figure 2.7). Resolutions adopted accumulated CoP after CoP until there were many of them. Therefore, at CoP8 in 1992 it was decided to undertake a review of all adopted resolutions with a view to repealing those, or parts of those, that were defunct and to consolidating remaining resolutions, or parts of resolutions, that dealt with the same subject in order to remove parts that were conflicting or duplicative.[38] Since this review, some of the resolutions that remained have been revised at subsequent CoPs, others deleted when they were superseded by a more significant revision or were considered to be outdated. In total, up until CoP20, 354 Resolutions have been adopted although only 100 remain in force, in their original or a revised form. This process of continually 'tidying up' resolutions has been copied by other biodiversity MEAs.

Initially, these resolutions contained all the agreements reached at meetings of the CoP, but at CoP9 in 1994 it was decided to also establish an additional record of 'Decisions' of the CoP. This was

a) to ensure that the Resolutions, as the 'soft law' of CITES, do not become cluttered with instructions or with recommendations that can be implemented by a single action and will then be out of date; and
b) to ensure that decisions that are made by the Conference but not included in Resolutions are readily accessible.

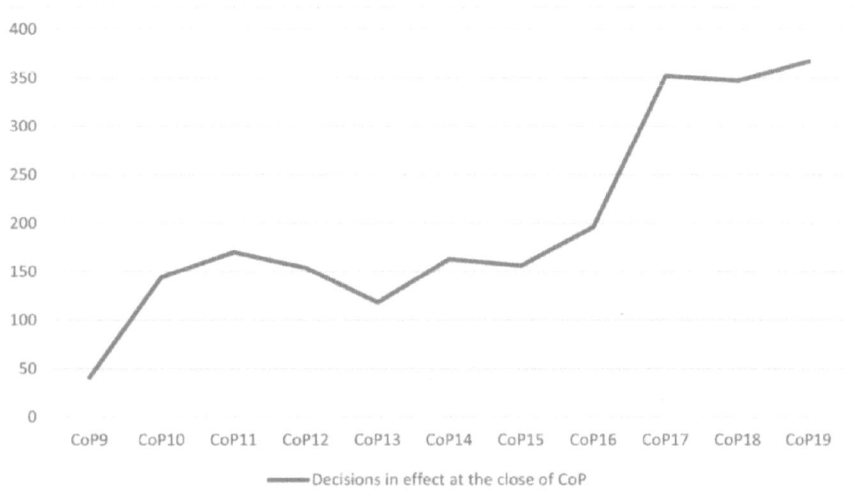

FIGURE 2.8 Number of decisions in effect at the close of each meeting of the Conference of the Parties.

In general, the decisions contain instructions or requests to permanent Committees, Working Groups, or the Secretariat. They are usually designed to remain in place for the period between consecutive meetings of the CoP. Although the format of the numbering system has changed a little over time, their number has grown significantly over the years (Figure 2.8).

2.4 INTERSESSIONAL COMMITTEES OF THE COP

The text of the Convention establishes a Conference of the Parties, but the nature of any other governing or advisory bodies is not specified, and this is left to the CoP to decide. From the outset, it was recognized by the Parties at the CoP that some sort of intersessional body or bodies were needed to oversee matters on behalf of the Parties between meetings of the CoP. Initially, those established were rather ad hoc with terms of reference, membership, and modus operandi sometimes unclear or not specified:

- Steering Committee—later Standing Committee
- Technical Expert Committee on Harmonization of Permit Forms and Procedures—later Technical Expert Committee and then Technical Committee
- Identification Manual Committee
- Committee on Standardized Taxonomy—later Nomenclature Committee
- Ten Year Review committees (Committees for each region, a Central Committee and Secretariat Committee)

Recognizing the urgent need to develop structures and procedures to ensure that meetings of the CoP operate in an efficient and effective manner. Canada, Switzerland, the UK, and Zimbabwe together drafted a detailed and ambitious proposal to reconcile these bodies and submitted it to CoP6 in 1987.[39] The Secretariat cautioned that the proposal, although welcome, had significant legal, financial, and practical shortcomings. Some commentators even suggested that the proposal might have been viewed as if government bureaucrats were trying to arrange for year-round international travel at no cost to themselves.[40] After detailed discussions, a Resolution was agreed which formalized, with great clarity, a permanent Standing Committee of the Conference of the Parties (the senior Committee), two science advisory committees (one for animals and one for plants), and Identification Manual and Nomenclature Committees. This sweeping change in the institutional structure was a decisive moment in the development of the Convention's administration. The Technical Committee was abolished, and the work of the Ten Year Review Committees was implicitly transferred to the Animals and Plants Committees under the term 'periodic review', with the Ten Years Review Committees themselves formally abolished only at CoP9 in 1994.[41]

2.5   STANDING COMMITTEE

At CoP1 in 1976, in view of the decision to hold an intersessional Special Working Session on Implementation Issues, a small provisional Steering Committee was established with a mandate to coordinate with the Secretariat, the organization of the Special Working Session, and CoP2 in 1979.[42] The five-Party Steering Committee (Canada, Ecuador, Ghana, Switzerland, and the USA) was chaired by Peter Gafner of Switzerland. It met three times: 21–22 February 1977 in Switzerland, 13–17 February 1978 in Costa Rica, and 9–10 October 1978 again in Switzerland. After it reported on the successful discharge of its mandate at CoP2, the Secretariat recommended that it be replaced by a 'Standing Committee of the Conference of the Parties'.[43] The Standing Committee's initial mandate was brief and rather general, but looked superficially similar to that of today:

i)   to provide guidance and advice to the Secretariat on the implementation of the Convention, on the preparation of meetings and on relations with the government of the host State of the Secretariat headquarters;
ii)   to act as bureau at meetings of the Conference of the Parties, in accordance with the rules of procedure; and
iii) to perform any other function as may be entrusted to it by the Conference of the Parties.[44]

The Standing Committee met for the first time in Bonn, Germany, on 22 June 1979.
   Out of concern about the cost implications, the initial composition of the Committee, to be nominated by the CoP, was of no more than nine Parties. To the

extent possible, these Parties were to include members from all geographical regions. Additionally, the Depositary Party (Switzerland) and previous and next host Party of the CoP were also to be members. The geographical regions described—Africa, Asia, Central and South America [later changed to Central and South America and the Caribbean], Europe, Oceania, and North America—were very different from those used by the United Nations. They arose from discussions involving the Secretariat and the Steering Committee chair at CoP2,[45] and the rationale for their delineation does not seem to have been recorded. The Standing Committee was given the authority to choose its own chair. One or two representatives from each member Party were permitted to attend meetings, and the Committee was also given the authority to invite observers to attend particular meetings or for particular meeting agenda items.

The concept of at least one voting Committee member from each CITES region remained only 'to the extent possible' until CoP9 in 1994 when the Committee's composition was changed to be fixed at one representative for regions with up to 15 Parties, two representatives for regions with 16 to 30 Parties, and three representatives for regions with more than 30 Parties.[46] However, at CoP11 in 2000 the number of representatives was changed to more closely align with the number of Parties in the region concerned:

- one representative for regions with up to 15 Parties
- two representatives for regions with 16 to 30 Parties
- three representatives for regions with 31 to 45 Parties
- four representatives for regions with more than 45 Parties

Nonetheless, as has been pointed out by some Parties at meetings of the CoP,[47] a democratic deficit can be said to remain in the current membership, as 13 Parties are required to result in one Standing Committee vote for the African region, whereas the same figure for North America is three (Table 2.4).

Right from its establishment at CoP2 in 1979, the term of office of regional members of the Standing Committee has lasted from the close of the CoP at which they were elected to expire at the close of the second regular CoP meeting thereafter—a period of around six years. This has proven a suitable time for members to become acquainted with their role. From CoP7 in 1989, alternate members, for each member, have also been chosen. These are only called into action in the unusual event that a member is unable to send any representative to a meeting. The role however also allows time for a mentorship or training period, with the expectation being that the alternate will become a full member after two inter-CoP intervals as an alternate member. As full member posts are highly prized, there have been very few instances of re-election of the same Party after their first term of office has been concluded (Table 2.5).

Although they had been informally invited to attend previously, the ability of Parties not members of the Standing Committee to send non-voting representatives to meetings of the Committee as observers was only enshrined in a resolution at

TABLE 2.4

Number of Parties in Each Region and Number of Members and Votes in the Standing Committee, Prior to CoP20 in 2025

Region	N° of Parties	N° of members/votes
Africa	53	4
Europe	51	4
Asia	38	3
Central and South America and the Caribbean	31	3
Oceania	9	1
North America	3	1
Depositary Party	–	1 (non-voting)
Previous CoP host Party	–	1 (non-voting)
Next CoP host Party	–	1 (non-voting)

CoP6 in 1987.[48] The Chair also had the authority to invite a representative of any observer organization to participate and some UN and intergovernmental organization did so. However, it was not until 2003, after the Committee's Rules of Procedure had been changed, that observers such as NGOs were first invited to attend and then only upon an invitation from the chair after consultation with the Committee members. Subsequently, their participation increased substantially, particularly as the importance of the work of the Committee developed (Table 2.6).

Although the Standing Committee's chair is a named Party, the direction of the meeting is often undertaken by a single person from the delegation of that Party (Table 2.7). The Committee's first chair was American Richard (Rick) Parsons, then head of the US Fish and Wildlife Service's Wildlife Permits Office (forerunner of the CITES Management Authority). Parsons became a CITES veteran and was the only person to attend all of the first 17 meetings of the Conference of the Parties (1976–2016). At CoP3 in 1981 in New Delhi, India, the host government was elected to chair the Standing Committee. This reflected the very active role that India played in the early days of the Convention,[49] a role that subsequently declined considerably. At the start, India was represented in the chair by Samar Singh, the only chair to serve three consecutive intersessional periods. Singh was an important figure in the development of nature conservation in India and at the time worked as director (Wildlife) in the Ministry of Environment and Forests. Later he went on to be Secretary-General of the World Wide Fund for Nature in India. Unable to attend CoP6 in 1987, he was replaced by Mr M.K. Ranjitsinh, who was also a key figure in the development of nature conservation in India. After this long tenure by India, chairs were subsequently elected only for a single intersessional period until CoP11 in 2000, after which they remained for two intersessional periods—the same duration as their Party's term of office on the Committee. All of the subsequent chairs have been senior CITES officials in their

TABLE 2.5
Regional Members (and Alternates Shown in Brackets) of the Standing Committee after the Close of Each Meeting of the Conference of the Parties

Meeting of the CoP	CoP2 (1979)	CoP3 (1981)	CoP4 (1983)	CoP5 (1985)	CoP6 (1987)	CoP7 (1989)	CoP8 (1992)	CoP9 (1994)	CoP10 (1997)
Africa 1	Zaire	Zaire	Kenya	Kenya	Malawi	Malawi (Morocco)	Senegal (Namibia)	Namibia (South Africa)	Namibia (South Africa)
Africa 2								Senegal (Burkina Faso)	Burkina Faso (Tunisia)
Africa 3								Sudan (Tanzania)	Sudan (Tanzania)
Africa 4									
Asia 1	Nepal	Nepal	India	India	Nepal	Nepal (Malaysia)	Thailand (India)	Japan (Pakistan)	Japan (Pakistan)
Asia 2								Thailand (India)	Saudi Arabia (Indonesia)
CSAC 1	Brazil	Brazil	Costa Rica	Costa Rica	Peru	Peru (not known)	Trinidad and Tobago (Panama)	Argentina (Panama)	Argentina (Ecuador)
CSAC 2								Trinidad and Tobago (St Lucia)	Panama (St Lucia)
CSAC 3									
Europe 1	UK	France	France	Germany	UK	Sweden (Denmark)	France (not known)	UK (France)	UK (France)
Europe 2								Russian Federation (Bulgaria)	Russian Federation (Bulgaria)
Europe 3									Italy (Czech Republic)
Europe 4									
N. America	USA	Canada	Canada	USA	USA	Canada (USA)	Canada (USA)	Mexico (Canada)	Mexico (USA)
Oceania	Australia	Papua New Guinea	Papua New Guinea	Australia	Australia	New Zealand (Papua New Guinea)	Papua New Guinea (New Zealand)	Papua New Guinea (New Zealand)	Papua New Guinea (New Zealand)

(Continued)

TABLE 2.5 (Continued)

Regional Members (and Alternates Shown in Brackets) of the Standing Committee after the Close of Each Meeting of the Conference of the Parties

Meeting of the CoP	CoP11 (2000)	CoP12 (2002)	CoP13 (2004)	CoP14 (2007)	CoP15 (2010)	CoP16 (2013)	CoP17 (2016)	CoP18 (2019)	CoP19 (2022)
Africa 1	Malawi (Morocco)	Senegal (Namibia)	Cameroon (Guinea)	Democratic Republic of Congo (Mali)	Democratic Republic of Congo (Mali)	Niger (Congo)	Niger (Morocco)	Morocco (Senegal)	Chad (Democratic Republic of the Congo)
Africa 2	Burkina Faso (Cameroon)	Cameroon (Guinea)	Zambia (Botswana)	Zambia (Botswana)	Botswana (Namibia)	Botswana (Namibia)	Namibia (Madagascar)	Namibia (Madagascar)	Madagascar (Zambia)
Africa 3	Tanzania (Kenya)	Tanzania (Kenya)	Kenya (Uganda)	Kenya (Uganda)	Uganda (Ethiopia)	Uganda (Ethiopia)	Ethiopia (Kenya)	Ethiopia (Kenya)	Kenya (United Republic of Tanzania)
Africa 4	Tunisia (Ghana)	Tunisia (Ghana)	Ghana (Egypt)	Ghana (Egypt)	Egypt (Niger)	Egypt (Madagascar)	Congo (Chad)	Congo (Chad)	Morocco (Senegal)
Asia 1	Nepal (Malaysia)	Thailand (India)	China (India)	China (India)	Japan (Jordan)	Japan (Jordan)	China (Japan)	China (Japan)	Japan (China)
Asia 2	Saudi Arabia (Indonesia)	Malaysia (United Arab Emirates)	Malaysia (Jordan)	Iran (Pakistan)	Iran (Pakistan)	Indonesia (China)	Indonesia (Nepal)	Indonesia (Nepal)	Indonesia (India)
Asia 3	Peru	Trinidad and Tobago (Panama)	Japan (United Arab Emirates)	Japan (Jordan)	Kuwait (Saudi Arabia)	Kuwait (Saudi Arabia)	Kuwait (Republic of Korea)	Kuwait (Republic of Korea)	Kuwait (Singapore)
CSAC 1	Panama (Nicaragua)	Nicaragua (Costa Rica)	Chile (Colombia)	Chile (Colombia)	Colombia (Peru)	Colombia (Peru)	Peru (Brazil)	Peru (Brazil)	Brazil (Ecuador)
CSAC 2	St. Lucia (St Vincent and the Grenadines)	St. Lucia (St Vincent and the Grenadines)	Nicaragua (Costa Rica)	Costa Rica (Guatemala)	Costa Rica (Guatemala)	Guatemala (Honduras)	Guatemala (Honduras)	Honduras (Nicaragua)	Honduras (Nicaragua)

TABLE 2.5 *(Continued)*

Regional Members (and Alternates Shown in Brackets) of the Standing Committee after the Close of Each Meeting of the Conference of the Parties

Meeting of the CoP	CoP11 (2000)	CoP12 (2002)	CoP13 (2004)	CoP14 (2007)	CoP15 (2010)	CoP16 (2013)	CoP17 (2016)	CoP18 (2019)	CoP19 (2022)
CSAC 3	Sweden (Denmark)	Sweden (Denmark)	Saint Vincent and the Grenadines (Dominica)	Saint Vincent and the Grenadines (Dominica)	Dominica (Bahamas)	Dominica (Bahamas)	Bahamas (Dominican Republic)	Bahamas (Dominican Republic)	Dominican Republic (Cuba)
Europe 1	Norway (Turkey)	Norway (Turkey)	Czech Republic (Bulgaria)	Bulgaria (Italy)	Bulgaria (Italy)	Hungary (Poland)	Hungary (Poland)	Poland (Ireland)	Poland (Ireland)
Europe 2	Italy (Czech Republic)	Germany (UK)	Iceland (Portugal)	Iceland (Norway)	Norway (Belgium)	Norway (Israel)	Israel (Albania)	Israel (Belarus)	UK (Belarus)
Europe 3		Germany (UK)	Germany (UK)	UK (Czech Republic)	UK (Czech Republic)	Portugal (Belgium)	Portugal (Belgium)	Belgium (Spain)	Belgium (Spain)
Europe 4					Ukraine (Russian Federation)	Ukraine (Russian Federation)	Russian Federation (Georgia)	Russian Federation (Georgia)	Georgia (Israel)
N. America	Canada (USA)	Canada (USA)	Mexico (Canada)	Mexico (USA)	USA (Canada)	USA (Canada)	Canada (Mexico)	Canada (Mexico)	USA (Canada)
Oceania	New Zealand (Papua New Guinea)	New Zealand (Papua New Guinea)	Papua New Guinea (New Zealand)	Papua New Guinea (New Zealand)	Australia (Vanuatu)	Australia (Vanuatu)	Australia (Fiji)	Australia (Fiji)	New Zealand (Australia)

TABLE 2.6
Meetings, Dates, Venues, and Participants (Where Available) at Meetings of the Standing Committee

Standing Committee meeting, year, and venue	Total participants
SC1 (1979) Bonn, Germany	Not located
SC2 (1980) Bonn, Germany	Not located
SC3 (1980) Nairobi, Kenya	Not located
SC4 (1981) New Delhi, India	–
SC5 (1981) Gland, Switzerland	Not located
SC6 (1981) Christchurch, New Zealand	Not located
SC7 (1982) Gland, Switzerland	Not located
SC8 (1983) Gaborone, Botswana	–
SC9 (1983) Gaborone, Botswana	–
SC10 (1983) Gland, Switzerland	Not located
SC11 1984) Gland, Switzerland	Not located
SC12 (1985) Buenos Aires, Argentina	–
SC13 (1985) Lausanne, Switzerland	Not located
SC14 (1986) Ottawa, Canada	–
SC15 (1987) Ottawa, Canada	–
SC16 (1987) Ottawa, Canada	Not located
SC17 (1988) San José, Costa Rica	Not located
SC18 (1988) Lausanne, Switzerland	Not located
SC19 (1989) Lausanne, Switzerland	–
SC20 (1989) Lausanne, Switzerland	–
SC21 (1990) Lausanne, Switzerland	Not located
SC22 (1990) Nairobi, Kenya	Not located
SC23 (1991) Lausanne, Switzerland	Not located
SC24 (1992) Lausanne, Switzerland	Not located
SC25 (1992) Kyoto, Japan	–
SC26 (1992) Kyoto, Japan	–
SC27 (1992) Kyoto, Japan	–
SC28 (1992) Lausanne, Switzerland	30
SC29 (1993) Washington, DC, USA	57
SC30 (1993) Brussels, Belgium	50
SC31 (1994) Geneva, Switzerland	84
SC32 (1994) Fort Lauderdale, USA	–
SC33 (1994) Fort Lauderdale, USA	–
SC34 (1994) Fort Lauderdale, USA	–
SC35 (1995) Geneva, Switzerland	72
SC36 (1996) Geneva, Switzerland	82
SC37 (1996) Rome, Italy	82
SC38 (1997) Harare, Zimbabwe	–
SC39 (1997) Harare, Zimbabwe	–
SC40 (1998) London, UK	112

TABLE 2.6 *(Continued)*
Meetings, Dates, Venues, and Participants (Where Available) at Meetings of the Standing Committee

Standing Committee meeting, year, and venue	Total participants
SC41 (1999) Geneva, Switzerland	158
SC42 (1999) Lisbon, Portugal	101
SC43 (2000) Gigiri, Kenya	–
SC44 (2000) Gigiri, Kenya	–
SC45 (2001) Paris, France	120
SC46 (2002) Geneva, Switzerland	164
SC47 (2002) Santiago, Chile	–
SC48 (2002) Santiago, Chile	–
SC49 (2003) Geneva, Switzerland	
SC50 (2004) Geneva, Switzerland	208
SC51 (2004) Bangkok, Thailand	–
SC52 (2004) Bangkok, Thailand	–
SC53 (2005) Geneva, Switzerland	185
SC54 (2006) Geneva, Switzerland	257
SC55 (2007) The Hague, the Netherlands	–
SC56 (2007) The Hague, the Netherlands	–
SC57 (2008) Geneva, Switzerland	255
SC58 (2009) Geneva, Switzerland	247
SC59 (2010) Doha, Qatar	–
SC60 (2010) Doha, Qatar	–
SC61 (2011) Geneva, Switzerland	259
SC62 (2012) Geneva, Switzerland	313
SC63 (2013) Bangkok, Thailand	–
SC64 (2013) Bangkok, Thailand	–
SC65 (2014) Geneva, Switzerland	374
SC66 (2016) Geneva, Switzerland	531
SC67 (2016) Johannesburg, South Africa	–
SC68 (2016) Johannesburg, South Africa	–
SC69 (2017) Geneva, Switzerland	564
SC70 (2018) Sochi, Russian Federation	498
SC71 (2019) Geneva, Switzerland	–
SC72 (2019) Geneva, Switzerland	–
SC73 (2021) Online	826
SC74 (2022) Lyon, France	369
SC75 (2022) Panama City, Panama	–
SC76 (2022) Panama City, Panama	–
SC77 (2023) Geneva, Switzerland	558
SC78 (2025) Geneva, Switzerland	594

TABLE 2.7

Chairs of the Standing Committee and the Person Leading for That Delegation

CoP	Standing Committee
CoP2–CoP3	USA (Mr Richard Parsons)
CoP3–CoP4	India (Mr Samar Singh)
CoP4–CoP5	India (Mr Samar Singh)
CoP5–CoP6	India (Mr Samar Singh/Mr M.K. Ranjitsính)
CoP6–CoP7	USA (Mr Ralph Mogenweck)
CoP7–CoP8	Malawi (Mr Matthew Matemba/Mr Henri Nsanjama)
CoP8–CoP9	New Zealand (Mr Murray Hosking)
CoP9–CoP10	Japan (Mr Nobutoshi Akao)
CoP10–CoP11	UK (Mr Robert Hepworth)
CoP11–CoP12	USA (Mr Kenneth Stansell)
CoP12–CoP13	USA (Mr Kenneth Stansell)
CoP13–CoP14	Chile (Mr Cristían Maquieira Astaburuaga)
CoP14–CoP15	Chile (Mr Cristían Maquieira Astaburuaga)
CoP15–CoP16	Norway (Mr Øystein Størkersen)
CoP16–CoP17	Norway (Mr Øystein Størkersen)
CoP17–CoP18	Canada (Ms Carolina Caceres)
CoP18–CoP19	Canada (Ms Carolina Caceres)
CoP19–CoP20	USA (Ms Rosemarie Gnam/Ms Naimah Aziz)

home country with the exception of Chile's Cristían Maquieira Astaburuaga, who was a career diplomat. At its sixty-eighth meeting in 2016, Carolina Caceres of Canada became the first woman to chair the Standing Committee (Figure 2.9), completing a unique treble as she had previously chaired Committee I at CoP16 in 2013 and also the Animals Committee. Her tenure at the helm of the Standing Committee was a particular triumph, successfully delivering an unprecedented workload for the Committee and chairing its first ever online meeting (due to COVID-19 restrictions) in May 2021.

The Standing Committee normally holds two 'regular' meetings between meetings of the CoP and 'supplementary' meetings at the CoP venue immediately before and immediately after the CoP. Regular meetings initially lasted one day, but momentum quickly built, and from 2004 each regular meeting has lasted a full five-day week. Formerly, meetings were often hosted by individual Parties, but this practice fell out of favour in 2001, until revived in 2018 when SC70 was held in Sochi, Russian Federation. Otherwise, regular meetings have been held in Switzerland, where meeting facilities are subsidized by the Depositary Government. COVID-19 restrictions led to one meeting being held online in May 2021, and when face-to-face meetings resumed, France hosted in the city of Lyon in March 2022.

FIGURE 2.9 Carolina Caceres of Canada chaired not only the Standing Committee 2016–2022 but also the Animals Committee, and Committee I at CoP16 in 2013.

Photo by IISD/ENB | Kiara Worth

The mandate of the Standing Committee was clarified at CoP3, and voting procedures clarified at CoP4. Further refinements have been made to the mandate, composition, and functioning of the Standing Committee at regular intervals until the present day. Amongst the most significant of these have been

- increased mandate to cover oversight of the Secretariat's budget and fundraising;
- assignment of lead role for CITES compliance measures;
- establishment and oversight of a Memorandum of Understanding between the Committee and the Executive Director of the UNEP;
- provision of funding to enable members from developing Parties to attend meetings; and
- the designation of an alternate members in the event of the inability of a member to attend a meeting.

The Committee's current mandate is as follows:

a) undertake the tasks directed to it by the Conference of the Parties, including those related to the handling of general and specific compliance matters;
b) provide general policy and general operational direction to the Secretariat concerning the implementation of the Convention;
c) provide guidance and advice to the Secretariat on the preparation of agendas and other requirements of meetings, and on any other matters brought to its attention by the Secretariat in the exercise of its function;

d) oversee, on behalf of the Parties, the development and execution of the Secretariat's budget as derived from the Trust Fund and other sources, and all aspects of fund raising undertaken by the Secretariat in order to carry out specific functions authorized by the Conference of the Parties, and to oversee expenditures of such fundraising activities;

e) implement and, as needed, review and revise the Memorandum of Understanding (MoU) between the Standing Committee and the Executive Director of the United Nations Environment Programme;

f) provide coordination and advice as required to the other committees established by this Resolution and provide direction to and coordination of its working groups and subcommittees;

g) carry out, between one meeting of the Conference of the Parties and the next, such interim activities on behalf of the Conference of the Parties as may be necessary, including advising on emerging operational or policy issues identified by Parties or the Secretariat until direction on the matter is provided by the Conference of the Parties;

h) draft resolutions or decisions for consideration by the Conference of the Parties;

i) report to the Conference of the Parties on the activities it has carried out between meetings of the Conference;

j) act as the Bureau at meetings of the Conference of the Parties, until such time as the Bureau of the Conference of Parties for the specific meeting has been constituted; and

k) perform any other functions as may be entrusted to it by the Conference of the Parties.

Although superficially similar, the duties of the Committee are a good deal more elaborate than in their original form when it started in 1979. There is more emphasis on oversight of the Secretariat's activities, a legacy of previous tensions between some of the Parties and the Secretariat. A notable new task since its early days is in the field of financial oversight for which it has established a Finance and Budget Subcommittee. Through the Subcommittee, the Standing Committee supervises the Secretariat's expenditure both from the Trust Fund and externally raised funds. In addition to its general duties, the volume of work addressed to the Standing Committee by the CoP has increased relentlessly in recent years. The assignment in particular of lead responsibility for compliance matters is an increasingly complex and delicate task, which now takes up a considerable amount of the Committee's time.

Rather like the CoP, the Rules of Procedure of the Standing Committee aim for decisions to be taken by consensus as far as possible. If this is not the case, however, the Chair or a member may call for a vote. In this circumstance, decisions are made by a simple majority of the members casting an affirmative or negative vote. In the case of a tie, the proposal is considered as rejected unless the tie is broken by the vote of the Depositary Government. Voting is relatively rare in the Standing Committee.

2.6 TECHNICAL COMMITTEE

In addition to the Steering Committee (later the Standing Committee) established at CoP1 in 1976, at CoP2 in 1979 the Parties established a Technical Expert Committee on

Harmonization of Permit Forms and Procedure. The initial objective of the Committee was *to guide the progressive harmonization of permit forms and procedures.*[50] The Committee was also mandated to *deal with control over trade in Appendix II and III species.*[51] This included species being traded (legally), but in a manner detrimental to their survival and those being traded in contravention of national laws. The Committee held its first meeting in Bonn, Germany, June 1979, and its first chair was renowned Indian environmentalist and mountaineer Nalni Jayal. In his role as chair of the Committee, he then led one of the two main parallel in-session committees at CoP3 in 1981. At that CoP, the Committee's name was shortened to Technical Expert Committee and its mandate amended to[52]

- identify, by means of the continual review of the annual reports of the Parties and other techniques, problems with enforcement of the Convention and provide guidance to the Secretariat and the Parties on measures that may be undertaken to remedy these problems;
- review the implementation of the Convention by the Parties and make recommendations for harmonization of documents and procedures;
- draft resolutions for consideration by the Conference of the Parties; and
- perform any other functions as may be entrusted to it by the Conference of the Parties or by the Standing Committee.

Amongst the latter were responsibilities related to transport of live specimens,[53],[54] reviewing Parties' annual trade reports, and assisting with the preparation of a planned publication titled a Yearbook of International Wildlife Trade.[55] The Committee's membership was open to all Party experts. By CoP4 in 1983, the Committee was being chaired by the UK's John Goldsmith. Its influence had grown, such that its name was shortened again to Technical Committee,[56] and six regional co-ordinators selected by the CoP were appointed to assist its work. Its mandate expanded to cover what is now the Review of Significant Trade[57] and the control of tourist souvenir specimens[58] and of worked ivory.[59] With the chair passing to Germany's Gerhard Emonds at CoP5 in 1985, further work on trade in ranched and captive bred specimens, look-alike species,[60] and on quota setting[61] was referred to the Committee. The proposal at CoP6 in 1987 for restructuring the working of the Convention by Canada, Switzerland, the UK, and Zimbabwe[62] had envisaged the re-establishment of the Technical Committee, renamed as the Management Committee and with a very similar mandate as the Technical Committee. However, as the joint proposal from these Parties was rejected on grounds of cost and practicality, the then duties of the Technical Committee were disbursed between the Standing Committee and the newly established science committees for plants and animals. The Technical Committee was therefore disbanded.[63] The duty of the Technical Committee to form one of the sessional committees at meetings of the CoP was assigned to a new CoP 'Committee I' which dealt with scientific and technical issues.

2.7 ANIMALS AND PLANTS COMMITTEES

Plants were rather neglected during the early years of the implementation of CITES. At CoP4 in 1983, IUCN botanists submitted a report bemoaning the poor implementation of the Convention for plants, and they urged the establishment of a working group of botanical, conservation, and CITES experts to discuss in detail how the monitoring and control of international trade in endangered wild-collected plants could be made more effective under CITES.[64] CoP agreed that the Technical Committee establish a Plants Working Group subcommittee[65] to examine the implementation of the Convention for plants and to make recommendations to the CoP on how CITES could work better for plants. This subcommittee met four times: 27 February–3 March 1984 in Tucson, USA; April-May 1985 in the margins of CoP5; June 1986 in the margins of the meeting of the Technical Committee; and July 1987, in the margins of CoP6. A further two meetings in Lausanne, Switzerland (in the margins of CoP7), and in Kyoto, Japan (in the margins of CoP8), were later attributed to this group, having been initially described as meetings of the newly established Plants Committee.[66] The Plants Working Group was chaired by Bruce MacBryde, a botanist in the Scientific Authority of the USA. Its work led to the first CoP resolution on improving the regulation of trade in plants.[67]

In their joint proposal for the restructuring of the intersessional committees of CITES, Canada, Switzerland, the UK, and Zimbabwe suggested an overarching Scientific Committee with subcommittees for animals and plants. After considering the results of discussions in a working group at CoP6 in 1987, the CoP decided however to establish a Plants Committee and an Animals Committee in their own right, both reporting to the CoP.[68] The composition of the new Committees was initially comprised of one member from each of the CITES regions (Africa, Asia, Europe, North America, South and Central America and the Caribbean, and Oceania). At CoP9 in 1994 the Africa, Asia, and renamed Central and South America and the Caribbean regions were afforded two members to reflect their higher number of Parties, with the membership for Europe similarly raised to two at CoP11 in 2000 (Table 2.8).

Candidates for membership of the Animals and Plants Committees are proposed by Parties (latterly on the basis of CVs circulated in advance), but importantly they must be persons—a Party cannot propose themselves to be members and offer to provide the name of an expert later. The individual members are then selected from the candidate list by regional gatherings of Parties held in the margins of meetings of the CoP and later ratified by the CoP itself. Importantly, this means that they reflect the biogeographical fauna and flora of the region they come from and perhaps the scientific culture of their region, but they act in an individual capacity. Candidates are however often selected along geopolitical lines by the regional gatherings at the CoP, and in the case of some regions the selected candidate is often not physically present at the meeting of the CoP where they are elected. No account is taken either of how the specialism and expertise they might bring could contribute to the Committee's work more broadly. Despite these constraints, the Animals and Plants Committees have largely managed to steer clear of the politicization that has bedevilled the scientific advisory bodies of some other biodiversity conventions.[69]

The Committees were hampered at first by the lack of interpretation of their meetings into the working languages of the Convention, business being conducted in English only. Occasionally the Secretariat was able to obtain external funds to cover

TABLE 2.8

Members (and Alternates Shown in Brackets) of the Plants Committee at the Close of Each Meeting of the Conference of the Parties and Committee Meetings That Took Place during Their Tenure

Meeting of the CoP	CoP6 (1987)	CoP7 (1989)	CoP8 (1992)	CoP9 (1994)	CoP10 (1997)	CoP11 (2000)	CoP12 (2002)
PC meetings	1	2	3, 4, 5	6, 7, 8	9	10, 11, 12	13, 14
Africa 1	Seyani	Djieng	Kabuye (Moh)	Kabuye (Seyani)	Seyani (Donaldson)	Donaldson (Akpangana)	Donaldson (Akpangana)
Africa 2				Moh (Cunningham)	Moh (Ndabaneze)	Luke (not known)	Luke (Khalifa)
Asia 1	Shrestha	Shrestha	Sharma (Shidiki)	Shaari (Thitiprasert)	Shaari (Thitiprasert)	Shaari (Siswomartono)	Irawati (Shaari)
Asia 2				De-yuan (Siswomartono)	De-Yuan (Siswomartono)	Singh (Jia)	Singh (Jia)
CSAC 1	Mora de Retana	Mora de Retana	Reyna de Aguilar (Hernandez Camacho)	de Oliveira (Bascopé)	de Oliveira (Mereles)	Wekhoven (Mereles)	Mereles (Prasad)
CSAC 2				Hernandez Camacho (Mora de Retana)	Mora de Retana (Wekhoven)	Forero (Riviera Luther)	Forero (Riviera Luther)
Europe 1	Lucas	McGough	McGough (von Arx)	Clemente Muñoz (von Arx)	Clemente Muñoz (Supthut)	Clemente Muñoz (Werblan-Jakubiec)	Clemente Muñoz (Werblan-Jakubiec)
Europe 2					de Koning (Werblan-Jakubiec)	de Koning (Supthut)	Frenguelli (Lüthy)
N. America	MacBryde	MacBryde	MacBryde (Márquez Ramirez)	MacBryde (Márquez Ramirez)	von Arx (Llorens)	von Arx (Ramírez)	von Arx (Ramírez)
Oceania	Miller	Armstrong	Armstrong (Owen)	Armstrong (Owen)	Leach (Babao)	Leach (Gideon)	Leach (Gideon)
Nomenclature specialist							

(Continued)

TABLE 2.8 (Continued)

Members (and Alternates Shown in Brackets) of the Plants Committee at the Close of Each Meeting of the Conference of the Parties and Committee Meetings That Took Place during Their Tenure

Meeting of the CoP	CoP13 (2004)	CoP14 (2007)	CoP15 (2010)	CoP16 (2013)	CoP17 (2016)	CoP18 (2019)	CoP19 (2022)
PC meetings	15, 16	17, 18	19, 20	21, 22	23, 24	25	26, 27
Africa 1	Hafashimana (Akpagana)	Hafashimana (Akpagana)	Hafashimana (Luke)	Hafashimana (Luke)	Mahamane (Nghidinwa)	Mahamane (Nghidinwa)	Koumba Pambo (Lagarde)
Africa 2	Khayota (Luke)	Khayota (Luke)	Khayota (Akpagana)	Khayota (Akpagana)	Koumba Pambo (Khayota)	Koumba Pambo (Khayota)	Balama (Zanndouche)
Asia 1	Irawati (Shaari)	Partomihardjo (Zakaria)	Partomihardjo (Zakaria)	Fernando (Rahajoe)	Fernando (Rahajoe)	Atikah (Zeng)	Zeng (Tongdonae)
Asia 2	Thitiprasert (Sanjappa)	Thitiprasert (Sanjappa)	Zhihua (Al-Salem)	Zhihua (Al-Salem)	Lee (Al-Salem)	Lee (Al-Salem)	Atikah (Chong)
CSAC 1	Mereles (Mites Cadena)	Mites Cadena (Rivera Brusatin)	Mites Cadena (Rivera Brusatin)	Rauber Coradin (Ramírez)	Rauber Coradin (Ramírez)	Núñez Neyra (Olave)	Núñez Neyra (Olave)
CSAC 2	Riviera Luther (Proctor)	Riviera Luther (Mejía)	Riviera Luther (Richardson)	Riviera Luther (Albert Puentes)	Beltetón Chacón (Albert Puentes)	Beltetón Chacón (Albert Puentes)	Beltetón Chacón (Perez Camacho)
Europe 1	Clemente Muñoz (Debeljak)	Clemente-Muñoz (Debeljak)	Clemente-Muñoz (Kikodze)	Clemente-Muñoz (Kikodze)	Da Luz Carmo (Kikodze)	Da Luz Carmo (Kikodze)	Smyth (de Boer)
Europe 2	Frenguelli (Lüthy)	Sajeva (Lüthy)	Sajeva (Da Luz Carmo)	Sajeva (Da Luz Carmo)	Sajeva (Moser)	Moser (Wolf)	Moser (Wolf)
N. America	Gabel (Caceres)	Gabel (Sinclair)	Benitez Diaz (Sinclair)	Benitez Diaz (Sinclair)	Sinclair (Camarena Osorno)	Gnam (Benitez Díaz)	Lougheed (Benitez Díaz)
Oceania	Leach (Gideon)	Leach (Gideon)	Leach (Vuli Tuiwana)	Leach (Vuli Tuiwana)	Leach (Vuli Tuiwana)	Wrigley (Vuli Tuiwana)	Wrigley (Likiafu)
Nomenclature specialist		McGough	McGough	McGough	McGough	Klopper	Klopper

some interpretation costs, for instance from the government hosting a meeting, but it was only in 1999 that funding was made available from the CITES budget to provide full interpretation in all three working languages of the Convention for meetings of the Animals and Plants Committees. Until 2000, the committee members were not provided with financial support to attend meetings, but at CoP11 in that year the Secretariat was authorized to make provision in its budget for participation by members at meetings,[70] although at CoP14 in 2007, representatives coming from developed countries were precluded from accessing this support.[71]

Parties have always been able to be represented by observers at meetings of the Animals and Plants Committee. and the chair had the authority to invite representatives of intergovernmental organization (IGOs), NGOs, or other non-State observers. The latter's participation was not without controversy, particularly in the Animals Committee, which has always had a greater participation by conservation and animal welfare NGOs. After NGOs made particularly active interventions at an Animals Committee meeting in 1996 in the Czech Republic, members complained that this caused an imbalance in the working of the Committee, and there were concerns that NGOs may have unfairly influenced the debate on some issues.[72] Participation fees for observer organizations at science committee meetings had been set at 100 USD per participant, and one suggestion was to raise this to 200 USD and to establish a fund that could be used to help NGO participants from developing countries to attend, but this was not adopted. Nearly 30 years later, the standard participation charge for all observer organizations (other than those from the United Nations and its specialized agencies) attending meetings of the Standing, Animals, and Plants Committees remains at 100 USD for each observer participant.

The Plants Committee has always had a lower profile than the Animals Committee, with about half the number of participants at its meetings. Arguably the average level of expertise has been higher than at the Animals Committee, with participants more clubbable and discussions more collegiate and less confrontational. With one notable exception, participation at meetings of the Plants Committee has yet to pass 200, notably because observer organization participation is much lower than in the Animals Committee (Tables 2.9 and 2.10). The notable exception was the Committee's 25th meeting in 2021, which was held online due to COVID-19 restrictions. The participation of 527 at this meeting was 177% higher than any other and permitted engagement from many Parties who never normally attend Plants Committee meetings.

The Plants Committee was authorized to choose their own chair and vice-chair. Bruce MacBryde chaired its first meeting in London, UK, from 2–5 November 1988, continuing his earlier role as chair of its 'predecessor', the Plants Working Group subcommittee (of the former Technical Committee). He was succeeded by Jim Armstrong, a larger-than-life Australian who resigned in mid-term after being appointed Deputy Secretary-General at the Convention's Secretariat. His replacement, Margarita Clemente Muñoz, an academic at the University of Cordoba in Spain, was to become the dominate figure in the history of the CITES Plants Committee (Figure 2.10). Whilst her forceful personality and professorial style was not to everyone's liking, she single-handedly raised the profile of the Plants Committee and plants in CITES generally. She had boundless energy and enthusiasm and legendary focus. At a Plants Committee meeting in Veracruz, Mexico, in 2014, proceedings

TABLE 2.9
Meetings, Dates, Venues, and Participants at Meetings of the Animals Committee

Animals Committee meeting, year and venue	IGO and NGO observer participants	Total participants
AC1 (1988) Bern, Switzerland	6	20
AC2 (1989) Montevideo, Uruguay	6	20
AC3 (1989) Lausanne, Switzerland	6	22
AC4 (1990) Darwin, Australia	12	35
AC5 (1991) Vancouver, Canada	10	30
AC7 (1992) Kyoto, Japan (at CoP8)	-	-
AC8 (1992) Harare, Zimbabwe	16	32
AC9 (1993) Brussels, Belgium	27	43
AC10 (1994) Beijing, China	19	38
AC11 (1994) Fort Lauderdale, USA	28	57
AC12 (1995) Antigua, Guatemala	31	73
AC13 (1996) Průhonice, Czech Republic	51	108
AC14 (1998) Caracas, Venezuela	58	126
AC15 (1999) Antananarivo, Madagascar	42	131
AC16 (2000) Shepherdstown, USA	82	140
AC17 (2001) Hanoi, Vietnam	46	118
AC18 (2002) San José, Costa Rica	47	115
AC19 (2003) Geneva, Switzerland	50	134
AC20 (2004) Johannesburg, South Africa	46	130
AC21 (2005) Geneva, Switzerland	38	121
AC22 (2006) Lima, Peru	52	154
AC23 (2008) Geneva, Switzerland	56	162
AC24 (2009) Geneva, Switzerland	42	145
AC25 (2011) Geneva, Switzerland	58	174
AC26 (2012) Geneva, Switzerland	70	168
AC27 (2014) Veracruz, Mexico	81	189
AC28 (2015) Tel Aviv, Israel	78	178
AC29 (2017) Geneva, Switzerland	104	256
AC30 (2018) Geneva, Switzerland	151	326
AC31 (2021) Online	233	756
AC32 (2023) Geneva, Switzerland	104	263
AC33 (2024) Geneva, Switzerland	124	328

were interrupted by a small earthquake which visibly alarmed some delegates not familiar with such things. Undeterred, she pressed on with business until persuaded that it may be wise to seek security advice from the hosting government before continuing. During her tenure she also chaired Committee I at CoP11 in 2000 and initiated and ran a master's course on Management, Conservation and Control of Species

TABLE 2.10

Meetings, Dates, Venues, and Participants at Meetings of the Plants Committee

Plants Committee meeting, year and venue	IGO and NGO observer participants	Total participants
PC1 (1988) London, UK	14	31
PC2 (1991) Zomba, Malawi	5	20
PC3 (1992) Chang Mai, Thailand	8	26
PC4 (1993) Brussels, Belgium	12	31
PC5 (1994) San Miguel, Mexico	28	58
PC6 (1995) Tenerife, Spain	25	72
PC7 (1996) San José, Costa Rica	32	80
PC8 (1997) Pucón, Chile	16	77
PC9 (1999) Darwin, Australia	14	65
PC10 (2000) Shepherdstown, USA	14	56
PC11 (2001) Langkawi, Malaysia	10	67
PC12 (2002) Leiden, the Netherlands	15	64
PC13 (2003) Geneva, Switzerland	15	69
PC14 (2004) Windhoek, Namibia	12	68
PC15 (2005) Geneva, Switzerland	13	81
PC16 (2006) Lima, Peru	20	93
PC17 (2008) Geneva, Switzerland	25	106
PC18 (2009) Buenos Aires, Argentina	12	101
PC19 (2011) Geneva, Switzerland	21	121
PC20 (2012) Dublin, Ireland	25	111
PC21 (2014) Veracruz, Mexico	16	99
PC22 (2015) Tbilisi, Georgia	21	108
PC23 (2017) Geneva, Switzerland	41	152
PC24 (2018) Geneva, Switzerland	44	172
PC25 (2021) Online	104	527
PC26 (2023) Geneva, Switzerland	38	151
PC27 (2024) Geneva, Switzerland	43	190

Subject to International Trade at the International University of Andalucía. After 20 years at the helm of the Plants Committee, even she became tired. One of her successors became the first CITES science committee chair to come from Africa. Aurélie Flore Koumba Pambo was born in the Republic of Congo but represented Gabon at CITES meetings and coincidentally is a graduate from Clemente Muñoz's master's course. Her initiation to the post of chair was challenging, to say the least, with her first meeting online due to COVID-19 restrictions. She has subsequently managed to combine her Plants Committee responsibilities with high profile diplomatic roles on behalf of Gabon at the United Nations in New York and latterly as a co-facilitator at the Congo Basin Forest Partnership (Table 2.11).

TABLE 2.11
Chairs of the Plants Committee

Intersessional period	Chair
CoP6–CoP7	Mr Bruce MacBryde (North America)
CoP7–CoP8	Mr Bruce MacBryde (North America)
CoP8–CoP9	Mr Jim Armstrong (Oceania)
CoP9–CoP10	Mr Jim Armstrong (Oceania)/Ms Margarita Clemente Muñoz (Europe)
CoP10–CoP11	Ms Margarita Clemente Muñoz (Europe)
CoP11–CoP12	Ms Margarita Clemente Muñoz (Europe)
CoP12–CoP13	Ms Margarita Clemente Muñoz (Europe)
CoP13–CoP14	Ms Margarita Clemente Muñoz (Europe)
CoP14–CoP15	Ms Margarita Clemente Muñoz (Europe)
CoP15–CoP16	Ms Margarita Clemente Muñoz (Europe)
CoP16–CoP17	Ms Margarita Clemente Muñoz (Europe)
CoP17–CoP18	Ms Adrianne Sinclair (North America)
CoP18–CoP19	Ms Aurélie Flore Koumba Pambo (Africa)
CoP19–CoP20	Ms Aurélie Flore Koumba Pambo (Africa)

FIGURE 2.10 Margarita Clemente Muñoz of Spain, chair 1997–2016 and doyen of the Plants Committee.

Photo courtesy of Diario Cordoba S.A.U.

The mandate of the Plants Committee has remained remarkably similar since its inception even if some of its language has been improved and brought up-to-date with emerging terminology and Convention procedures. The Committee works under, and in accordance with the instructions of, the Conference of the Parties and currently has the following key tasks:

- Conduct an ongoing review of trade in Appendix II plant species, identify instances of unsustainable exports of Appendix II plants species and suggest remedial measures to address this [the Review of Significant Trade]
- Conduct a review of the placement of plant species in the Appendices [the Periodic Review]
- Advise on plant nomenclature
- Advise on tools to identify specimens of CITES-listed plant species
- Provide plant-related advice to Parties on request
- Carry out any other instruction decided by the Conference of the Parties

When established, the Committee also had a remit to assist and advise Parties in the preparation of publicity material for Convention listed plants, but this is no longer specifically mentioned in its terms of reference. On the other hand, it now has a more significant role in the provision of advice on the making of non-detriment findings and export quotas for plant species. The initial mandate of the Plants Committee differed in one significant way from that of the Animals Committee, in that it included the, rather cryptic, instruction:

- serve, if so requested by the Conference of the Parties, as a plants working group

This relates back to the actions previously undertaken by the Technical Committee Plants Working Group subcommittee who prepared advice for the Conference of the Parties on proposals to amend the Appendices for plant species. The Plants Committee continued this role at CoP7 in 1989[73],[74] and in a more informal way at CoP8 in 1992,[75] but by CoP9 in 1994 the practice had stopped although the duty was still included in the Plants Committee terms of reference until CoP18 in 2019. Currently the Committee has a duty to provide advice related to scientific, technical, and nomenclatural aspects of proposals to amend the Appendices only if these are requested by Parties. The Committee though continues to take a close interest in amendments to the Appendices concerning plants, often as an advocate of listing plant species in the Appendices. By contrast, the Animals Committee has always been much more circumspect when it comes to amendments to the Appendices, largely because listing proposals related to animals are often more 'politically' charged than those for plants. In this regard, neither of the CITES scientific committees however has as an instruction or role as strong as the Scientific Council of CITES' sister convention, the Bonn Convention on the Conservation of Migratory Species. The Bonn Convention Scientific Council is given the specific task of assessing proposals for the amendment of the Convention's Appendices from a scientific and technical standpoint and providing advice to its Conference of the Parties—a role undertaken by the Secretariat in CITES.

The Plants Committee always had a responsibility to perform any other functions related to plants that may be entrusted to it by the Conference of the Parties. Initially, these did not greatly exceed the specific responsibilities listed in its terms of reference, but latterly the number of ad hoc instructions from the CoP to the Plants Committee has increased significantly.

TABLE 2.12

Members (and Alternates Shown in Brackets) of the Animals Committee at the Close of Each Meeting of the Conference of the Parties and Committee Meetings That Took Place during Their Tenure

Meeting of the CoP	CoP6 (1987)	CoP7 (1989)	CoP8 (1992)	CoP9 (1994)	CoP10 (1997)	CoP11 (2000)	CoP12 (2002)
AC meetings	1, 2, 3	4, 5, 6	7, 8, 9, 10	11, 12, 13	14, 15	16, 17. 18	19, 20
Africa 1	Olindo	Olindo (Mankoto Mbaelele)	Hutton (Ngong Nje)	Hutton (Severre)	Cameroon (Zimbabwe)	Howell (Chidziya)	Chidziya (Mahmoud)
Africa 2				Ngog Nje (Won Wa Musiti)	Tanzania (Togo)	Griffin (Bagine)	Griffin (Bagine)
Asia 1	Clark	Wang Sung (Ishii)	Ishii (Wang Sung)	Soehartono (Seneviratne)	Soehartono (Seneviratne)	Soehartono (Hussain)	Pourkazemi (Ishii)
Asia 2				Giam (Hussain)	Giam (Hussain)	Tunhikorn (Giam)	Tunhikorn (Giam)
CSAC 1	Villalba-Macias	Villalba-Macias (Vincente Rodrigues)	Inchaustegui (Quero de Peña)	Quero de Peña (Ramos Tangarona)	Quero de Peña (Ramos Tangarona)	Incháustegui (Ramos Tangarona)	Incháustegui (Ramos Tangarona)
CSAC 2				Francisco Lara (Inchaustegui)	Francisco Lara (Inchaustegui)	Micheletti (Ojeda)	Micheletti (Ojeda)
Europe 1	Dollinger	Blanke (Morgan)	Blanke (Morgan)	Blanke (Kucera)	Hoogmoed (Tew)	Hoogmoed (Althaus)	Althaus (Ibero Solana)
Europe 2					Rodics (Bistrom)	Rodics (Fleming)	Rodics (Sorokin)
N. America	Shoesmith	Shoesmith (Dane)	Medina Gonzalez (Dauphiné)	Dauphiné (Salgado y Bonilla)	Lieberman (Dauphiné)	Lieberman (Medellin)	Medellin (Johnson)
Oceania	Richmond	Jenkins (Bani)	Jenkins (Bani)	Jenkins (Hay)	Jenkins (Hay)	Hay (Walting)	Hay (Walting)
Nomenclature specialist							

Meeting of the CoP	CoP13 (2004)	CoP14 (2007)	CoP15 (2010)	CoP16 (2013)	CoP17 (2016)	CoP18 (2019)	CoP19 (2022)
PC meetings	21, 22	23,24	25, 26	27, 28	29, 30	31	32, 33
Africa 1	Chidziya (Mahmoud)	Zahzah (Mahmoud)	Zahzah (Mahmoud)	Madzikanda (Fouda)	Mensah (Fouda)	Mensah (Fouda)	Ngalié (Wakibara)
Africa 2	Bagine (Zahzah)	Bagine (Maurihungirire)	Kasiki (Maurihungirire)	Kasiki (Kalema-Zikusoka)	Mukasa (Kalema-Zikusoka)	Kalema-Zikusoka (Maha)	Kasoma (Diouck)
Asia 1	Pourkazemi (Ishii)	Pourkazemi (Ishii)	Pourkazemi (Ishii)	Khamdan (Ishii)	Khamdan (Ishii)	Mobaraki (Terada)	Mobaraki (Diesmos)
Asia 2	Prijono (Giam)	Prijono (Giam)	Suharsono (Giam)	Suharsono (Giam)	Giyanto (Mobaraki)	Giyanto (Desmos)	Hamidy (Terada)
CSAC 1	Jolon Morales (Calvar Agrelo)	Calvar Agrelo (Estrada Andino)	Calvar Agrelo (Estrada Andino)	Calvar Agrelo (Herrerra)	Calvar Agrelo (Herrerra)	Ramadori (Ramirez)	Ramadori (Ramirez)
CSAC 2	Vogel (Velasco Barbieri)	Álvarez Lemus (Velasco Barbieri)	Álvarez Lemus (Ouboter)	Ouboter (Álvarez Lemus)	Ouboter (Álvarez Lemus)	Gongora (Suazo Díaz)	Gongora (Suazo Diaz)
Europe 1	Althaus (Ibero Solana)	Althaus (Ó Críodáin)	Fleming Lortscher	Fleming (Gaynor)	Fleming (Gaynor)	Ziková (Benyr)	Lörtscher (Novitsky)
Europe 2	Rodics (Sorokin)	Ibero Solana (Suciu)	Ibero Solana (Suciu)	Lörtscher (Nemtzov)	Lörtscher (Nemtzov)	Lörtscher (Novitsky)	Ziková (Benyr)
N. America	Medellín (Gabel)	Medellín (Gnam)	Caceres Gnam	Caceres (Gnam)	Gnam (Caceres)	Benitez Díaz (Lougheed)	Benitez Díaz (Gnam)
Oceania	Hay (not known)	Hay (Aruga)	Robertson (Hay)	Robertson (Hay)	Robertson (Makan)	Robertson (McIntyre)	Robertson (McIntyre)
Nomenclature specialist		Grimm	Grimm	Grimm	van Dijk	van Dijk	van Dijk

The Committee is obliged to provide a report to each meeting of the Conference of the Parties on the work it has undertaken.

Whilst the Animals Committee's remit is very similar to that of the Plants Committee and it has the same structure and composition, its profile has been much higher, due to the more media-friendly nature of animals, NGO interest in its business, and the resultant contentiousness of some of its work. This is reflected in the size of its meetings, which, after an initial establishment period, have consistently been over 50% higher than those of the Plants Committee. Recent meetings of the 12-member committee have had about 300 observer participants (from Parties, IGOs, NGOs, and others), who play a vital role in the committee delivering its mandate.[76]

The chairing of the Animals Committee has been dominated by persons from the European and North American regions (Table 2.13). The chair of its first meeting held between 7–11 November 1988 in Berne, Switzerland, was Canadian Merlin Shoesmith, an affable biologist from the Department of Natural Resources in the province of Manitoba. He proved a 'safe pair of hands' whilst the Committee was finding its feet. His successor, Hank Jenkins (Figure 2.11), was a very different character. He was a biologist who was head of the national CITES Management Authority of Australia housed in the Australian National Parks and Wildlife Service.[77] His inimitable dress sense and sometimes irreverent sense of humour masked a deep knowledge of his subject, and he advanced the work of the Committee considerably, particularly with regard to procedures for the Review of Significant Trade. He was subsequently elected chair of a CoP working group reviewing the criteria for the amendment of

---

TABLE 2.13
Chairs of the Animals Committee

Intersessional period	Chair
CoP6–CoP7	Mr Merlin Shoesmith (North America)
CoP7–CoP8	Mr Merlin Shoesmith (North America)
CoP8–CoP9	Mr Robert (Hank) Jenkins (Oceania)
CoP9–CoP10	Mr Robert (Hank) Jenkins (Oceania)
CoP10–CoP11	Mr Robert (Hank) Jenkins (Oceania)
CoP11–CoP12	Mr Marinus Hoogmoed (Europe)
CoP12–CoP13	Mr Thomas Althaus (Europe)
CoP13–CoP14	Mr Thomas Althaus (Europe)
CoP14–CoP15	Mr Thomas Althaus (Europe)
CoP15–CoP16	Mr Carlos Ibero Solana (Europe)
CoP16–CoP17	Ms Carolina Caceres (North America)
CoP17–CoP18	Mr Mathias Lörtscher (Europe)
CoP18–CoP19	Mr Mathias Lörtscher (Europe)
CoP19–CoP20	Mr Mathias Lörtscher (Europe)

FIGURE 2.11   Robert (Hank) Jenkins of Australia, chair of the Animals Committee 1992–2000. He typified the sort of knowledgeable and colourful characters who made CITES what it is.

Photo by IISD/ENB | Kiara Worth

Appendices I and II. The chairing of the Animals Committee in recent years has been dominated by two Swiss vets: Thomas Althaus and Mathias Lörtscher, each having headed the Swiss CITES Management Authority. Each served for three intersessional periods as chair (a combined total of 12 meetings over nearly 18 years). Both were characterized by their calm and neutrality and for their defence of the role of the Committee and its scientific independence. Between their terms of office, Canada's Carolina Caceres was appointed the first female chair of the Committee, but she had to relinquish the role after one intersessional period as she was chosen to chair the Standing Committee.

At CoP13 in 2004, Australia expressed concern about the increasing cost of the meetings of the Animals, Plants, and the then existing Nomenclature Committees and the fact that in its view, recommendations from the scientific committees were not attracting consistent support from the Conference of the Parties. Consequently, they believed that the arrangements for the committees could benefit from a review to give the Parties confidence that their structure was the best available and that monies allocated for the committees were justified.[78] The CoP agreed to undertake a review with the objective of improving and facilitating the performance of the functions of the Committees.[79] The Committees themselves were charged with drafting terms of reference for the review which was overseen by the Standing Committee.

The Standing Committee was presented with some facts and figures about the science committees from the Secretariat[80] and a self-evaluation by the science committees themselves[81] and came to the conclusion that the scientific committees achieved a generally high level of performance on the high-priority tasks assigned to them and often with very limited resources or a reliance on voluntary effort. The Standing Committee noted the limited budgetary resources for the science committees and that care should be taken to ensure that the tasks assigned to them by the CoP were within their mandate. The principal change proposed by the Standing Committee was the abolition of the Nomenclature Committee and the incorporation of its work into the mandates of the Plants and Animals Committees.[82] This was approved by CoP14 in 2007. As a possible cost-saving measure, the Secretariat had suggested the possible merger of the Animals and Plants Committees at CoP12 in 2002,[83] and they repeated the suggestion in the context of the review of the science committees, the results of which were considered at CoP14 in 2007.[84] The idea of further considering this option however brought widespread opposition, not least from the chairs of the science committees themselves.[85] Whist the review of the science committees was a useful stock check; it could probably have been made more valuable if it had been undertaken by a body external to the CITES family. However, this would have required financial resources which have never been available at sufficient levels in CITES. One further suggestion arising from the review of the science committees was only implemented at CoP15 in 2010 and was that the Chair of each of the science committees should be replaced in his/her role of regional member by his/her alternate. This freed up the time of the chair to concentrate on running the meetings and intersessional business of their committee which, given the increase in their workloads, was a welcome development.

Some matters discussed by the meetings of the CoP are common to both the Animals and Plants Committees. The revision of the criteria for listing species in the Appendices being one of the more fundamental issues. In order to prepare coherent scientific advice on the question of the listing criteria, the Animals and Plants Committees met jointly in Brussels, Belgium, in 1993 and in Shepherdstown, USA, in 2000. These meetings were considered a success and led to regular joint sessions being held with meetings of the science commitees being held back-to-back in 2005, 2006, 2008, 2012, 2014, 2017, 2018, and 2024.

At CoP14 in 2007, the Secretariat suggested that because members of the Animals and Plants Committees are appointed in an individual capacity, rather than as representatives of a Party, the Committees should adapt their Rules of Procedure in order to prevent and deal with any conflicts of interest relating to their activities.[86] In the interests of transparency, many other MEAs had adopted such measures. Although a number of Parties supported this suggestion,[87] when it was proposed to be incorporated into their Rules of Procedure at subsequent meetings of the science committees, they rebuffed the idea, considering it essentially an ethical matter and querying how a conflict of interest could be identified and who might make such a determination.[88] Consequently, at the following CoP, the Standing Committee was instructed to review this matter but, in 2011, pressed by some observer Parties and the Animals Committee Chair, decided that there was no need for the Rules of Procedure of the Animals and Plants Committees to address potential conflicts of interest of their members relating to their activities in the Committees.[89] After a particularly high-profile case involving a member of the Animals Committee, Denmark, on behalf of the European Union Member States, took up the charge.[90] Over the heads of the science committees, CoP16 in 2013 amended their terms of reference to include a definition of 'conflict of interest' and to provide provisions requiring candidate members for the science committee membership to disclose any current financial interest that might call into question his or her impartiality, objectivity, or independence in undertaking out their duties and for members to declare such any interest at the start of each meeting.[91] Even so, the latter declarations were initially made in private for the respective Committee to note in open session, and no science committee member has ever declared a conflict of interest, notwithstanding the existence of several circumstances where there may have been concerns. One problem is that in order to head off opposition from the science committees, the definition of 'conflict of interest' was limited to a current pecuniary interest and then only such interest which might significantly impair an individual's impartiality, objectivity, or independence in undertaking his or her duties as a member of a committee, with a candidate's employment not by itself constituting a conflict of interest. The handling of the potential conflict of interest issue by Animals and Plants Committees was not their finest moment. They failed to appreciate how their resistance to adopt simple guidelines to address this issue might be viewed by others.

The Animals and Plants Committees take decisions by consensus as far as possible. If this is not possible, the chair or members from at least two regions may call for a vote, and the matter is decided by a simple majority. As advisory bodies on scientific matters, votes in the Animals and Plants Committees have been exceptionally rare.

Despite considerable external pressures, the Animals and Plants Committees have largely succeeded in providing scientific advice to the CoP and the Parties which is both impartial and of creditable scientific quality.

2.8 NOMENCLATURE COMMITTEE

The importance of a common understanding between the Parties on the taxonomy and nomenclature of the species listed in the Appendices is explored elsewhere in this book. The CoP's Special Working Session on Implementation Issues in 1977 established a committee of experts under the Steering Committee to draft a provisional standard taxonomy and to compile a list of taxonomic bibliographic source materials.[92] This approach was formally adopted at CoP2 in 1979,[93] and the Ad Hoc Committee on Standardized Taxonomy, chaired by American herpetologist F. Wayne King was instructed to continue its work and expand the provisional list of standardized scientific species names of the vertebrates and of genera of plants and prepare a proposal to further expand and refine the standardized nomenclature of vertebrates and plants for presentation at CoP3. By CoP3, this group was being referred to as the Nomenclature Committee, but funding for its work was not forthcoming, and the

chair requested the CoP to decide whether or not the efforts at producing a standardized nomenclature should continue and, if so, requested Parties to assist in assur-
ing adequate funding for this work. The Committee continued its work nonetheless
and was also charged with providing guidance on the appropriate classification of
taxa and statistical data in the annual reports of the Parties.[94] At CoP4 in 1983 and
CoP5 in 1985, the Committee continued to report on progress in its activities even
if its status was still ad hoc, with no formal membership, solid funding, long-term
mandate, or reporting duties. At CoP4, F. Wayne King announced his retirement
and was replaced by Steve Edwards of the Association of Systematics Collections,
an American not-for-profit organization of institutions that had been instrumental
in producing some of the early standardized nomenclatures such as that for mammals. It was only at CoP6 in 1987, during the major restructuring of the Convention's
institutions, that the Nomenclature Committee was established on a more formal
footing.[95] The Committee was to have open membership, elect its own chair and
vice-chair, and report to the Conference of the Parties on its adopted mandate:

i)   develop and maintain nomenclatorial standard references for animals and
      plants, where necessary to the level of subspecies;
ii)   review the existing appendices with regard to the correct use of zoological
      and botanical nomenclature;
iii) upon request from the Secretariat, review proposals to amend the appendi-
      ces to ensure that correct names for the species and other taxa in question
      are used;
iv)   ensure that changes in nomenclature recommended by a Party do not alter
      the scope of protection of the taxon concerned; and
v)   make recommendations on nomenclature to the Conference of the Parties,
      other committees, working groups and the Secretariat.

Financial constraints remained however, and at CoP7 in 1989, Edwards threatened
to resign if funds were not provided for the work of the committee,[96] and these were
subsequently approved for the Committee to continue to operate. During the meeting,
the prevailing uncertainty resulted in the representatives of the Plants Committee
agreeing that the Nomenclature Committee should be dissolved and that responsibility for nomenclature be transferred to the Animals and Plants Committees[97]—this
did indeed occur, but not until 2007. In the meantime, the Nomenclature Committee
divided informally into two subcommittees with the nominal committee addressing
animals and the work relating to plant nomenclature being undertaken by the Plants
Committee.

Whilst an agreed taxonomy and nomenclature for the species covered by the
Convention would appear to be a fairly basic requirement for the smooth operation of the Convention, the task of assembling standard nomenclatural references
itself was not glamorous. The Nomenclature Committee depended on a handful of
dedicated experts whose work ensured that most species now have agreed scientific
names. The chairs of the Committee in particular laboured intently in dark corners
of conference centres over the years and rarely received the credit that their efforts
deserved (Table 2.14).

TABLE 2.14

Chairs of the Nomenclature Committee

Intersessional period	Chair
CoP2–CoP3	Mr F. Wayne King (North America)
CoP3–CoP6	Mr F. Wayne King (North America)
CoP4–CoP5	Mr Steve Edwards (North America)
CoP5–CoP6	Mr Steve Edwards (North America)
CoP6–CoP7	Mr Steve Edwards (North America)
CoP7–CoP8	Mr Steve Edwards (North America)—Fauna/ Mr Bruce MacBryde (North America)—Flora
CoP8–CoP9	Mr Steve Edwards (North America)—Fauna/ Mr Noel McGough (Europe)—Flora
CoP9–CoP10	Mr Steve Edwards (North America)—Fauna/ Mr Noel McGough (Europe)—Flora
CoP10–CoP11	Mr Marinus Hoogmoed (Europe)—Fauna/ Mr Noel McGough (Europe)—Flora
CoP11–CoP12	Mr Marinus Hoogmoed (Europe)—Fauna/ Mr Noel McGough (Europe)—Flora
CoP12–CoP13	Mr Marinus Hoogmoed (Europe)—Fauna/ Mr Noel McGough (Europe)—Flora
CoP13–CoP14	Ms Ute Grimm (Europe)—Fauna/ Mr Noel McGough (Europe)—Flora

As mentioned, at CoP14 in 2007, the Nomenclature Committee was abolished and its duties transferred to the Animals and Plants Committees, whose membership was augmented by a specialist on zoological or botanical nomenclature respectively. Ironically, as a regular part of the work of the Animals and Plants Committees, the establishment and maintenance of a standard species nomenclature now has higher profile than it did when it was presided over by a dedicated Committee.

2.9   IDENTIFICATION MANUAL COMMITTEE

A significant concern during the early years of the Convention was how customs officers in particular were going to identify specimens of CITES-listed species in trade. CoP1 in 1976 had decided that the development of identification aids to facilitate the enforcement of trade restrictions under the Convention was one of the issues to be discussed at the CoP's Special Working Session on Implementation Issues which was later held in 1977. At that meeting, the Steering Committee established a sub-committee to guide the preparation of an identification manual of the species included in the Appendices. The 17 experts in the sub-committee were drawn from Australia, Botswana (the Federal Republic of), Germany, India, Papua New Guinea, Switzerland, the UK, the USA, and Uruguay.[98] Switzerland chaired the sub-committee in the form of Peter Dollinger, who went on to become the only person to chair Committee I at a meeting of the CoP twice and was a key figure in the development of the Convention. CoP2 in 1979 noted the existence of the 'Committee of Experts' on this matter,[99] but it was not until CoP3 in 1981 that it was referred to as the Identification Manual Committee.[100] In addition to its core task, it was also charged at that meeting with providing guidance on the appropriate classification of taxa and statistical data in the annual reports of the Parties.[101] Its membership was open to any experts nominated by Parties, but its reporting obligations and rules of procedure were rather unclear. It met opportunistically in the margins of other CITES meetings. Over the years, other Parties provided experts for the Committee: Argentina, Austria, Canada, Colombia, Chile, Denmark, France, Guatemala, India, Italy, Sweden, but Switzerland (Peter Dollinger) was the driving force. Dollinger even opened a bank account in the Committee's name and successfully raised funds outside the official circuit to cover the costs of its activities. The Identification Manual project was given a boost when UNEP agreed to provide funding, which enabled the engagement of a part-time editor and a full-time secretary, work previously undertaken by Switzerland from 1979 to 1983. The mandate for the existence of the Identification Manual Committee was extended CoP after CoP until CoP6 in 1987 when it was given documented terms of reference and a modus operandi.[102] Ironically it was at this meeting that Dollinger resigned as chair, frustrated at the lack of financial support from Parties and their decision to add 500 species to the Appendices at the meeting.[103] No volunteers were found to replace him until ten years later at CoP10 in 1997 when Ms Ruth Landolt of Switzerland was elected to the role. However, subsequently only one Party expressed interest in appointing expert members to the Committee, and it was dissolved at CoP11 in 2000 with the duties regarding the Identification Manual transferred to the Secretariat.[104] Notwithstanding its somewhat disjointed history, the Committee had nonetheless overseen the production of 2082 identification datasheets, prepared by volunteer Parties (around half produced by Switzerland) to a standard format. Most datasheets had also been translated into French and Spanish with a Japanese version produced in 1986 and a Russian version in 1998. Thereafter, the Secretariat gradually converted all the Identification Manual datasheets into electronic format and made them available on the CITES website, a task completed in early 2007. However, these datasheets were not being revised to account for new information, and at CoP15 in 2010, in the face of a large number of new identification materials being made available, the Secretariat proposed converting the existing Identification Manual into a Web-based database with a partial 'wiki' format. After a review by the Animals, Plants, and Standing Committees, this process was completed, and at CoP19 in 2022, mention of an Identification Manual in CITES resolutions was removed completely in favour of 'materials for the identification of specimens of CITES-listed species'.[105]

2.10   TEN-YEAR REVIEW COMMITTEES

As part of a review of the species listed on the Appendices described in more detail in the species chapter of this book, a first complete review was to be undertaken over a ten-year period to ensure that species were correctly placed in the Appendices. At CoP3 in 1981 it was agreed that Regional Committees be established to review the trade and biological status of all Appendices I and II species indigenous to the region, with a Secretariat Committee reviewing such information for species not indigenous to any region. The Secretariat and regional Standing Committee members were charged with making the necessary arrangements for convening the meetings of the Regional Committees, and a Central Committee was established to coordinate the review.[106] There was a flurry of initial meetings of these committees:

> North American Region: Washington, USA, and Ottawa and Vancouver, Canada, in 1981 and 1982
> European Region: Paris, France, 26–27 January 1982
> Asian Region: Kathmandu, Nepal, 17–19 May 1982
> African Region: Nairobi, Kenya, 26–28 November 1982

No meetings of the Latin American or Oceania regions were held. The Secretariat Committee met in Cambridge, UK, on 16 December 1981, and the Central Committee met in Gland, Switzerland, 24–25 June 1982, where it elected Switzerland as chair.[107]

Although progress with the Ten-Year Review had been slower than anticipated, CoP4 in 1983 agreed that the prevailing arrangements should continue,[108] and these were noted at CoP5 in 1985.[109] In the absence of any physical meetings, the work was continued through the Standing Committee. By CoP6 in 1987, the newly established Animals and Plants Committees had been entrusted to continue the work undertaken by the Ten-Year Review Committees.[110] They had been charged with undertaking a 'periodic review' of species included in CITES Appendices by

- establishing a schedule for reviewing the biological and trade status of these species;
- identifying problems or potential problems concerning the biological status of species being traded; and
- informing the Parties of the need to review specific species, and assisting them in such reviews.[111]

The Animals and Plants Committees continued to prepare proposals under the Ten-Year Review banner however, and it was only at CoP9 in 1994 that the resolutions relating to the Ten-Year Review and its committees were formally repealed.[112] The mandate of the Ten-Year Review committees to review all Appendices I and II species was over-optimistic, and only a little over 100 proposals to delete species from the Appendices were generated by the committees—and not all of these were agreed by the CoP.

2.11   THE IMPORTANCE OF THE CONVENTION'S INTERSESSIONAL COMMITTEES

The intersessional committees of CITES have fulfilled a vital role in the development of the Convention's policy. Their restructuring at CoP6 in 1987, although developed on the spot at the meeting, proved to be an excellent model and remains largely unchanged today (Figure 2.12).

Most policy developments adopted at meetings of the CoP arise from work undertaken first in the intersessional committees, but not all Parties are willing or able to attend their meetings. Between 2013 and 2019, for instance, 73 Parties (40% of all

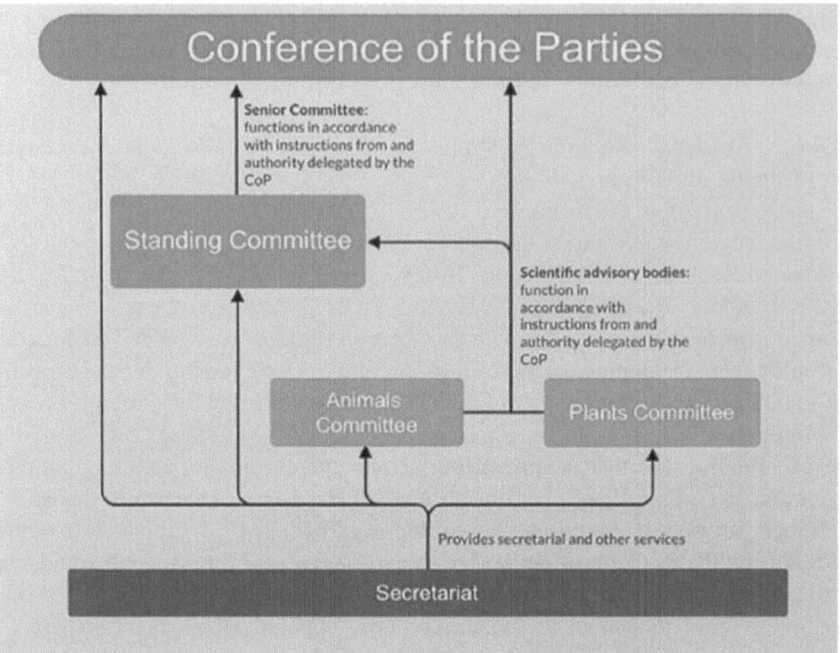

FIGURE 2.12   Administrative structure of CITES from 2007 to date.
Source: CITES website

Parties) did not attend any of the regular meetings of the Standing Committee—or any meetings of the science committees.[113] This may have been due to a lack of interest or lack of relevance of the issues discussed, but is also likely to be related to the cost of attending meetings, noting that unlike attendance at the meetings of the CoP, there is no scheme to provide financial support to assist developing country representatives to attend Standing Committee meetings. In some instances, all Parties are invited by Notification to the Parties to send comments on an issue to be discussed by the intersessional committees, but those Parties unable to attend meetings in person are greatly disadvantaged in terms of their influence on the recommendations made by the committees. Once developed and perfected at meetings of the intersessional committees, proposals sent to the CoP for approval are very difficult to change. Understandably, those who have worked long hours to negotiate a text that they are satisfied with are reluctant to reopen the matter when the document arrives at the meeting of the CoP.

NOTES

1 Anon (2022) *Espèces menacées (CITES). Département fédéral des affaires étrangères DFAE.* www.eda.admin.ch/eda/fr/dfae/politique-exterieure/droit-international-public/traites-internationaux/depositaire/esp%C3%A8ces-menacees-(cites).html (accessed 25.02.25).

2 Resolution Conf. 18.6.

3 Resolution Conf. 10.3.

4 Article VIII.6.

5 CITES Secretariat (undated) Reporting Requirements and Submission of Information. https://cites.org/sites/default/files/eng/resources/reporting/Reporting%20require-ments%20post%20CoP19.pdf (accessed 25.02.25).

6 Resolution Conf. 3.10.

7 Dec. 12.87.

8 CITES Secretariat (2022) *World Wildlife Trade Report*. Geneva, Switzerland.

9 UNEP-WCMC (undated) *CITES Wildlife TradeView*. https://tradeview.cites.org/ (accessed 25.02.25).

10 Resolution Conf. 11.3 (Rev. CoP19) para. 16 c.

11 IUCN (1976) *Report to the European Economic Community on Certain Aspects of the Protection of Species of Wild Fauna*. Unpublished report. 79pp. IUCN, Morges, Switzerland.

12 Plen. 1.3.

13 Plen. E. 2.1.

14 Council Regulation (EEC) No 3626/82 of 3 December 1982 on the implementation in the Community of the Convention on international trade in endangered species of wild fauna and flora and Commission Regulation (EEC) No 3418/83 of 28 November 1983 laying down provisions for the uniform issue and use of the documents required for the implementation in the Community of the Convention on international trade in endangered species of wild fauna and flora.

15 Resolution Conf. 6.5 and Resolution Conf. 8.2.

16 World Conservation Monitoring Centre and the IUCN Environmental Law Centre (1988) *Application of CITES in the European Economic Community*. 3 volumes. Unpublished Report to the Commission of the European Communities.

17 Council Regulation (EC) No 338/97 of 9 December 1996 on the protection of species of wild fauna and flora by regulating trade therein and Commission Regulation (EC) No 939/97 of 26 May 1997 laying down detailed rules concerning the implementation of Council Regulation (EC) No 338/97 on the protection of species of wild fauna and flora by regulating trade therein.

18 ASEAN (2004) *2004 ASEAN Statement on CITES on the Occasion of the Thirteenth Meeting of the Conference of the Parties Centre for International Law*. https://cil.nus.edu.sg/wp-content/uploads/2019/02/2004-ASEAN-Statement-on-CITES-1.pdf (accessed 25.02.25).

19 Resolution Conf 3.2.

20 SC66 Doc. 12.7.

21 Resolution Conf. 17.3.

22 CoP18 Inf. 67 (Rev. 1) and CoP19 Inf. 100 (Rev. 1).

23 Doc. 1.8 .

24 Resolution Conf. 1.7.

25 Secretariat (1978) *Proceedings of the Special Working Session of the Conference of the Parties*. Convention on International Trade in Endangered Species of Wild Fauna and Flora, Geneva, Switzerland, 17–28 October 1977. IUCN, Morges, Switzerland.

26 SC62 Summary Record. Item 10.1.

27 SC61 Doc. 11.1.

28 CITES Secretariat (2022) *Rules of Procedure of the Conference of the Parties* (as amended at the 19th meeting, Panama City, 2022). https://cites.org/sites/default/files/eng/cop/19/E19-CoP-Rules.pdf (accessed 25.02.25).

 * Not formally 'officers' of the CoP at the time of the CoP in question.

29 CITES (1986) *Proceedings of the Fifth Meeting of the Conference of the Parties.* Buenos Aires, Argentina, 22 April to 3 May 1985. Secretariat of the Convention. Lausanne, Switzerland, p. 151.

30 Doc. 6.3 (Rev.).

31 SC Doc. 31.4.8.

32 CoP16 Doc. 4.2 (Rev. 1).

33 Species Survival Network (2013) *Governments Strike a Blow Against Transparency.* https://ssn.org/press_releases/governments-strike-a-blow-against-transparency/ (accessed 25.02.25).

34 CITES World—Official Newsletter of the Parties. Issue Number 10—December 2002. CITES Secretariat, Geneva, Switzerland.

35 CITES Secretariat (1982) *Annual Report of the Secretariat 1981.* https://cites.org/sites/default/files/document/E-Annual%20Report%201981.pdf (accessed 25.02.25).

36 SC65 Doc. 21.

37 Proceedings of the First Meeting of the Conference of the Parties. Berne, Switzerland 2–6 November, 1976, p. 21.

38 Doc. 9.19.

39 Doc. 6.7.

40 Favre, D. (1989) *International Trade in Endangered Species. A Guide to CITES.* Martinus Nijhoff Publishers. Dordrecht, Boston and London.

41 Resolution Conf. 9.24.

42 Resolution Conf. 1.7.

43 Doc. 2.8.

44 Resolution Conf. 2.2.

45 Plen. 2.2 (Rev.).

46 Resolution Conf. 9.1.

47 CoP17 Plen. Rec. 4 (Rev. 1).

48 Resolution Conf. 6.1.

49 Jain, P. (2001) *CITES & India.* TRAFFIC-India, WWF-India and Ministry of Environment and Forests, New Delhi, India.

50 Resolution Conf. 2.5.

51 Resolution Conf. 2.6.

52 Resolution Conf. 3.5.

53 Resolution Conf. 3.16.

54 Resolution Conf. 3.17.

55 Resolution Conf. 3.10.

56 Resolution Conf. 4.4.

57 Resolution Conf. 4.7.

58 Resolution Conf. 4.12.

59 Resolution Conf. 4.14.

60 Resolution Conf. 5.16.

61 Resolution Conf. 5.21.

62 Doc. 6.7.

63 Resolution Conf. 6.1.

64 Doc. 4.17.

65 CoP4 Plen. 4.5. Proceedings page 109.

66 Armstrong, J. (Ed.) (1994) *CITES Plants Committee.* Report of Meeting. Brussels, Belgium, 6–8 September 1993. https://library.dbca.wa.gov.au/static/FullTextFiles/015815.pdf (accessed 25.02.25).

67 Resolution Conf. 5.14.

68 Resolution Conf. 6.1.

69 Koetz, T., Bridgewater, P., van den Hove, S. and Siebenhuener, B. (2008) The Role of the Subsidiary Body on Scientific, Technical and Technological Advice to the Convention on Biological Diversity as Science–Policy Interface. *Environmental Science & Policy*, 11: 505–516. https://doi.org/10.1016/j.envsci.2008.05.001.
70 Resolution Conf. 11.1.
71 Resolution Conf. 11.1 (Rev. CoP14).
72 AC13 Summary record.
73 Com. 7.1.
74 Com. 7.21.
75 Com.I 8.1 (Rev.).
76 CoP19 Doc. 24.
77 CITES Secretariat (2023) *CITES Pays Tribute to Dr Hank Jenkins*. https://cites.org/eng/news/tribute/dr-hank-jenkins (accessed 25.02.25).
78 CoP13 Doc. 11.1.
79 Decisions 13.9 and 13.10.
80 SC54 Inf. 4.
81 SC54 Inf. 5.
82 CoP14 Doc. 12.
83 CoP12 Doc. 9.1 (Rev. 1) Annex 4 (Rev. 2).
84 CoP14 Doc. 12.
85 CoP14 Com. II Rep. 2 (Rev. 1).
86 CoP14 Doc. 8.4.
87 CoP14 Com. II Rep. 4 (Rev. 1).
88 AC23 and PC17 Summary Records.
89 SC61 Summary Record.
90 CoP16 Doc. 11 (Rev. 1).
91 Resolution Conf. 11.1 (Rev. CoP16).
92 Document Conf. S.S. 1.7*.
93 Doc. 2.22 [and 2.22.1] adopted (see Plen. 2.9).
94 Resolution Conf. 3.10.
95 Resolution Conf. 6.1.
96 Doc. 7.17.
97 Com. 7.21.
98 Doc. 2.17.
99 Resolution Conf. 2.4.
100 Resolution Conf. 3.18.
101 Resolution Conf. 3.10.
102 Resolution Conf. 6.1.
103 Doc 7.16 Annex 1.
104 Resolution Conf. 11.19.
105 Resolution Conf. 19.4.
106 Resolution Conf. 3.20.
107 Doc. 4.37.
108 Resolution Conf. 4.26.
109 Plen. 5.9.
110 Doc. 7.41.
111 Resolution Conf. 6.1.
112 Resolution Conf. 9.24.
113 CoP19 Doc. 24.

3 CITES Secretariat

There are more than 1,000 multilateral environmental agreements (MEAs) in effect,[1] many of which have a secretariat. Most MEAs were established since 1970, and so the CITES Secretariat is amongst the earlier models. As such, it had little by way of precedent to follow in terms of organization and modus operandi.

The functions of the Secretariat are stated in some detail in Article XII of the Convention:

a) arrange for and service meetings of the Parties;
b) perform the functions entrusted to it in the Convention relating to the amendment of the Appendices;
c) undertake scientific and technical studies in accordance with programmes authorized by the CoP, including studies concerning standards for appropriate preparation and shipment of living specimens and the means of identifying specimens;
d) study the reports of Parties and request from Parties such further information with respect thereto as it deems necessary to ensure implementation of the Convention;
e) invite the attention of the Parties to any matter pertaining to the aims of the Convention;
f) publish periodically and distribute to the Parties current editions of Appendices I, II and III together with any information which will facilitate identification of specimens of species included in those Appendices;
g) prepare annual reports to the Parties on its work and on the implementation of the Convention and such other reports as meetings of the Parties may request;
h) make recommendations for the implementation of the aims and provisions of the Convention, including the exchange of information of a scientific or technical nature;
i) perform any other function as may be entrusted to it by the Parties.

For the most part, these are tightly constrained and under the authority of the Parties, although the Secretariat does have the ability to take a more proactive role in making any recommendations for the implementation of the aims and provisions of the Convention to the Parties. Over time the Parties have delegated a considerable number of other responsibilities to the Secretariat. After 50 years, these other tasks have become rather overwhelming and have put the Secretariat and its staff under increasing pressure.

3.1 ESTABLISHMENT

IUCN played a decisive role in the drafting of CITES, and there may have been an expectation that it therefore would have been given the task of providing the Convention's secretariat. However, arising from the Stockholm Conference on the Human Environment in June 1972, the United Nations General Assembly established an

DOI: 10.1201/9781003542278-3

environmental management body in December 1972,[2] which was later named the United Nations Environment Programme (UNEP). Although not a United Nations treaty, environmental conventions like CITES often have their Secretariats administered by UNEP. According to IUCN's Wolfgang Burhenne,[3] he was contacted by newly appointed first UNEP Executive Director Maurice Strong, during the negotiations at the plenipotentiary conference of 12 February to 2 March 1973, with a request that UNEP should provide the secretariat for CITES. Strong was not amongst the delegation of UNEP at the conference, but his intervention was enough for the final text of the convention to state that

> Upon entry into force of the present Convention, a Secretariat shall be provided by the Executive Director of the United Nations Environment Programme. To the extent and in the manner he considers appropriate, he may be assisted by suitable inter-governmental or non-governmental international or national agencies and bodies technically qualified in protection, conservation and management of wild fauna and flora.

This rather ambiguous wording created some tensions between UNEP and the CITES Parties which remain today. One immediate problem is that the Executive Director of the UNEP does not have the authority to provide such a secretariat; that power rests with the UNEP Governing Council (now the UN Environment Assembly). The Final Act of the plenipotentiary conference in 1973 included a Resolution expressing the hope that the UNEP Governing Council would approve the undertaking of secretariat functions by UNEP. It gave UNEP until 1 September 1973 to act, after which the Depositary Government of Switzerland was requested to assume the secretariat responsibilities on an interim basis pending consideration of the matter at the first Conference of Contracting States.[4] In the event, the very first Governing Council of UNEP authorized the Executive Director to provide secretariat services for the new convention.[5] Subsequently, and in line with the delegation possibility provided for in the Convention text, UNEP contracted IUCN to provide secretariat services from 1 April 1974—before the entry into force of the Convention. This arrangement continued, through a series of contracts, until 31 August 1984. Initially, CITES work was undertaken by IUCN staff who also had other duties within IUCN,[6] but on 1 April 1975 Jaques Berney became the first staff member to be hired by IUCN specifically to work on CITES. On 15 April 1975, shortly after the tenth State deposited its instrument of ratification meaning that the Convention could come into force, he began work as 'executive secretary' of the Secretariat.

3.2 TIMELINE

In late 1975, Maurice Strong was replaced by Mustafa Tolba as Executive Director of UNEP, and by CoP1 in 1976 there were already signs of concern about the size of the Secretariat amongst the Parties. Far from the one staff member in post, Parties considered that 11 staff with the following functional responsibilities were the minimum required.[7]

Executive Secretary	1
Combined Scientific, Technical, Statistical, and Legal Officers	5
Administrator	1
Secretaries	3
Translators	1

A Resolution at CoP1 urged the Executive Director of UNEP to provide extra Secretariat capabilities based on this analysis, and the UNEP Governing Council in May 1977 endorsed the recommendations.[8] By 1 January 1977, Berney had been joined by a secretary from the central IUCN staff and 'a young Argentinian biologist [Obdulio Menghi] who dealt with all material in Spanish'.[9] Menghi went on to become the Secretariat's first head of science.

By March 1978, UNEP's approach to funding the CITES Secretariat had changed. The Executive Director proposed a cost-sharing arrangement including direct financial support from the Parties for the operation of the Secretariat. This approach was endorsed at the UNEP Governing Council in May 1978, which agreed to fund the holding of CoP2 but no further CoP meetings. Arriving on 1 May 1978, in the middle of this fast-moving change of circumstances, was the CITES Secretariat's first Secretary-General, Peter H. Sand. Sand was a widely travelled German lawyer who was recruited from FAO, where he had been a senior legal officer.[10] Later in 1978, UNEP proposed that its funding of the CITES Secretariat should be gradually reduced and cease, preferably by the end of 1982.[11] Faced with this situation, decisive action was needed by the Parties. At CoP2 in March 1979 a majority of Parties agreed that an amendment to the text of the Convention was needed to specifically permit the CoP to adopt financial provisions, in particular to run the Secretariat. An extraordinary CoP was held in Bonn, Germany, on 22 June 1979 where this was agreed. Prior to this, at CoP2, interim arrangements regarding the financial participation of Parties to the running of the Secretariat were adopted until entry into force of the amendment to the Convention text—after two-thirds of the Parties who were members at the time had deposited their formal instruments of acceptance with the Depositary Government. These interim arrangements were to last some time as the amendment to the text of the Convention only entered into force in April 1987. Parties agreed to ask UNEP to establish a trust fund, with agreed administration terms of reference, into which each Party could pay a contribution based on an adjusted version of the scale of contributions already established for running the United Nations. In April 1979, the UNEP Governing Council agreed to establish such a trust fund. The Secretariat was given day-to-day responsibility for managing the CITES Trust Fund under the direction of the CoP and, between meetings of the CoPs, under the supervision of the Standing Committee.

At CoP2 in 1979, the permanent staff of the Secretariat consisted of three professional officers and two secretaries.[12] This was as planned in the 1980–81 budget. This budget also envisaged funds for translators and consultants to work on trade monitoring and statistics, technical assistance missions, development of guidelines for legislation and administration, preparation of an identification manual and taxonomy.[13] Notably this permanent staffing was much smaller than CoP1 had agreed was necessary.

During the first few years of the implementation of the Convention, given the lack of experience and resources, the Secretariat focussed its activities on core functions. These included reminding Parties of their obligations, organizing meetings of the Parties, dealing with requests to include species in Appendix III, and sending out news in the form of 'Notification[s] to the Parties' which began with N° 1 on 18 August 1975 (Figure 3.1).

The Secretariat did however also establish a list of English common names of many of the species listed in the Appendices and a country-by-country listing of species mentioned in Appendices I and II. By CoP2 in 1979, the Secretariat had also started to build relations with relevant partners: the International Whaling Commission, the Customs Co-operation Council (now World Customs Organization), the International Plant Protection Convention over model phytosanitary certificates, and the International Air Transport Association over the transport of live specimens.[14] Cooperation with other organizations went on to become a significant feature of the Secretariat's work. Early consultancy arrangements were also put into place. As part of their contract from UNEP, IUCN was asked to review aspects of the trade reported by Parties, and this was undertaken by a newly formed specialist group of IUCN's Survival Service Commission named the TRAFFIC Group (Trade Records Analysis of Flora and Fauna in Commerce). Their report on the International Trade in Felidae 1977 was submitted to Parties and was the first analysis of Parties' annual

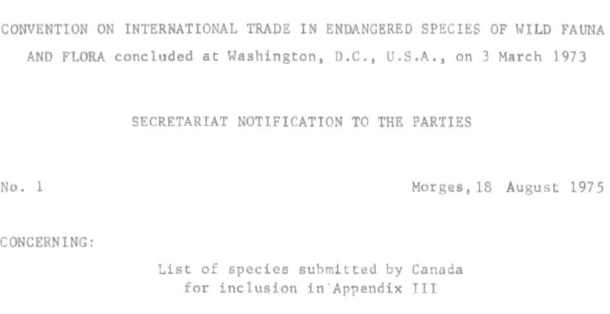

FIGURE 3.1 The first ever Notification to the Parties.

CANADA

FAUNA

MAMMALIA

CETACEA
Monodontidae   Monodon monocerus
Eschrichtidae   Eschrichtius robustus (glaucus)
Balaenopteridae   Balaenoptera musculus
                          Megaptera novaengliae
Balaenidae   Balaena mysticetus
                          Eubalaena spp.

CARNIVORA
Ursidae   Ursus (Thalarctos) maritimus
Mustelidae   Martes americana atrata
Felidae   Felis concolor cougar

PINNIPEDIA
Odobenidae   Odobenus rosmarus

ARTIODACTYLA
Bovidae   Bison bison athabascae

AVES

FALCONIFORMES
Accipitridae   Aquila chrysaetos
                          Haliaetus leucocephalus alascanus
Falconidae   Spp. Ø

Pandionidae   Pandion haliaetus
STRIGIFORMES
Strigidae   Nyctae scandiaea
                          Strix nebulosa

FIGURE 3.1   (Continued)

trade reports.[15] The Secretariat and World Wildlife Fund (now the World Wide Fund for Nature) also commissioned the IUCN Environmental Law Centre to produce an Index of Species Mentioned in Legislation, which was a computerized index of species mentioned in national and international legislation and used information from the biennial reports of Parties on legislative, regulatory, and administrative measures taken to enforce the Convention. By 1980, the Secretariat had established a directory containing the contact details for management and scientific authorities in each Party.[16] Peter Sand left his post of Secretary-General to become assistant director-general of IUCN on 1 June 1981, and the Standing Committee asked Jaques Berney to cover the post whilst a new Secretary-General was recruited. That did not occur until April 1982 when a Canadian lawyer, Eugène Lapointe, began his eventful term of office. Lapointe previously led a legislative unit in the Canadian Ministry, the Department of Industry, Trade, and Commerce and in that capacity had been responsible for dealing with a number of international agreements including CITES.[17] He inherited a Secretariat that was composed of three professional officers and two support staff—the same level as in May 1978, although the number of Parties and resulting workload had greatly increased since then. The gradual phasing out of UNEP funding for the provision of the Secretariat was concurrent with discussions over a host country agreement for the Secretariat which would provide certain tax and other advantages. Switzerland was interested in concluding such a host-country agreement with IUCN, which would include the Secretariat of CITES. However (the Federal Republic of) Germany, Kenya, and the UK also expressed interest in hosting the Secretariat. After discussions with these Parties, it was concluded that as neither the Secretariat, the Standing Committee, nor the CoP constituted an international organization in a legal sense, it was not possible to conclude a headquarters treaty with Germany, Switzerland, or the UK and that moving to Kenya may prove too expensive in the short-term.[18] A resolution was passed at CoP3 in 1981 confirming that the headquarters of the Convention Secretariat should remain in Switzerland, in association with the headquarters of IUCN. In an opening speech at the same meeting, IUCN Director General Lee Talbot, who had been a key player in the creation of CITES, described the secretariat arrangement between IUCN and CITES as eminently satisfactory. IUCN handled the administrative matters for the CITES unit under the joint project arrangement with UNEP and the CITES Secretariat interacted directly with the Parties and in all practical terms operated as an independent entity. The Secretariat had been housed in IUCN headquarters in Morges, Switzerland, initially in a converted hotel named Les Uttins (Figure 3.2) and later in the nearby and more spacious Floréal building.

In 1980, IUCN moved to Gland, further west along Lake Geneva, and the CITES Secretariat went with it. The move coincided with a change in the relationship between IUCN and WWF. The latter took over finance, personnel, and administrative responsibilities for IUCN and hence the CITES Secretariat.[19] This caused some friction with the Secretariat, and in November 1983 the Secretariat drew the attention of the Standing Committee to the necessity of re-evaluating the administration of the Secretariat as they believed that the arrangement involving IUCN no longer satisfied the needs of the rapidly evolving Convention.[20] A working group of the Standing Committee was established, and the Secretariat contracted a consultant to assist the working group with its activities. By July 1984, the working group was able to recommend to the Standing Committee that the Executive Director of UNEP should exercise more direct control over the Secretariat and that the staff at the time should be moved from IUCN to UNEP. The Executive Director of UNEP was immediately informed of, and agreed to, this request. On 1 November 1984, less than a year after the matter was first formally raised, the Secretariat moved from IUCN and to a new premises in a rather anonymous office block in nearby Lausanne. Moving from IUCN to UNEP meant that the host country agreement issue was resolved as the Secretariat was covered by the headquarters agreement applying to all United Nations bodies in Switzerland. This change in circumstances is notable not only for

FIGURE 3.2   Les Uttins in Morges, Switzerland, which was the first accommodation of the CITES Secretariat.

Photo courtesy of Fritz Vollmar

its speed but also for the fact that it was initiated largely by the Secretariat itself and was organized by the Standing Committee and not the CoP.

The savings provided by the move allowed the appointment of an extra support staff member. The Executive Director of UNEP also agreed to allow funds from the 13% overhead retained by UNEP from Party contributions to the CITES Trust Fund to be used to hire an Administrative Assistant. Japan became the first Party to second a staff member to the Secretariat when Yoshio Kaneko joined for two years as Special Projects Co-ordinator in 1985.[21] As the hiring of staff depended on the availability of funding and many Parties were not prompt in paying their contributions to the Trust Fund, regular staff were employed on short contracts which were continually renewed. The Secretariat at last looked to be assembling a staff complement which reflected its increasing obligations, although with an increasing reliance on external donor funding, rather than the CITES Trust Fund, to achieve this. In 1984, the Secretariat produced the first official publicity brochure about CITES in English French and Spanish. It was sent to Parties, NGOs, and airlines. Security stamps to affix to permits to confirm their authenticity were distributed to those Parties wishing to use them, after the idea was first agreed at CoP3 in 1984. The Secretariat also began a series of regional training seminars to build capacity and cooperation. At CoP5 in 1985 the Parties instructed the Secretariat to embark on a structured review of shortcomings in the implementation of the Convention by Parties in line with Article XIII. This 'Review of alleged infractions' was to become a significant activity until replaced by other compliance procedures. It was the start of the CITES compliance regime described in more detail later in this book.

By early 1986, the Secretariat had 13 permanent staff members (six professionals and seven support staff) and two long-term consultants (one professional and one support). Staff were divided into a small finance/administration team, teams dealing with scientific operations, technical operations, and an Ivory Control Unit.[22] However, less than half of these staff posts were funded from the regular budget allocated by the Conference of the Parties. External donor funding was also required to hire 14 short-term consultants to work on scientific studies and trade monitoring, implementation, and enforcement.

The Secretariat noted that external donor funding can be obtained and properly used only when the basic operations of the Secretariat were securely funded and, despite its achievements, warned of 'the dark looming cloud of financial deficits and shortage of funds to effect programmes in the future'.[23] A new workstream, which started in 1986, was the ivory export quota system that had been agreed at CoP5 in 1985.[24] Increasing concerns about the level of trade in African elephant ivory made this an imperative. The establishment of an annual ivory export quota by each state having a population of African elephants and wishing to export raw ivory, and the registration of existing stockpiles, was a major undertaking required by the Parties. As a result of these new measures, the Secretariat appointed a full-time co-ordinator in a newly established Ivory Control Unit. The problem, as so often was the case, was that this work was dependent on funds being found by the Secretariat to put this into practice. The estimated cost was 100,000 USD. Parties were slow to respond, and so initially over three quarters of the funds required were received from the Ivory Division of the Japan General Merchandise Importers Association (JGMIA). This industry involvement left Joe Yovino, head of the Ivory Control Unit, uncomfortable, and in early 1988 he resigned as a result.[25] To avoid the impression of a conflict of interest, the funds from JGMIA were subsequently channelled through the World Wide Fund for Nature.[26]

In addition to their regular duties, senior Secretariat staff were also assigned the role of regional co-ordinators to provide a focal point for Parties in their region. This was designed to try and bring the Secretariat closer to the Parties. It doubtless had that effect but also risked the development of 'fiefdoms' within the Secretariat. Such an effect had been noted at other multi-lateral environment agreement secretariats, complicating staff management. In 1989 and also in 1992, the Secretariat suggested that it should establish regional officers, stationed in UNEP offices in Africa, Asia/Oceania, and Central and South America and the Caribbean,[27] but this idea was not followed up for financial reasons.

By mid-1989, the Secretariat had been reorganized, in particular following two additional staff secondments. Environmental activist Jean-Patrick Le Duc, formerly of the Natural History Museum in Paris, was seconded by France as enforcement and information officer. He was later taken on as a staff member, serving in various roles in the Secretariat before leaving in controversial circumstances in 1998. The other secondee in 1989 was David Brackett from the Department of Renewable Resources in the Government of the Northwest Territories in Canada. He stayed until 1991 as management co-ordinator and after returning to Canada retained a keen interest in CITES, subsequently chairing Committee I at CoP10 in 1997. Further second-ments from Parties followed in the first quarter of 1990: the Secretariat's first Plants Officer, Ger Van Vliet from the Netherlands, and John Gavitt, formerly a special agent with the US Fish and Wildlife Service, to work on enforcement.[28] Brackett's arrival in particular led to the Secretariat being reorganized into two Units—science and management, which reflected the structure of CITES authorities in the Parties.

However, unease in some quarters about the conduct of the Secretariat was grow-ing. The Secretary-General told CoP7 in 1989 that the Secretariat efforts had been frustrated by the fact that it had been subject to considerable criticism.[29] The Standing Committee had called for a long term or strategic plan for its operations, the first of which was submitted at that CoP.[30] In the eyes of some, the annual ivory export quota system in which the Secretary-General had invested considerable effort was not working, and African elephants were considered by some to be moving into the category of 'endangered' under CITES. At CoP7 in October of 1989 proposals were made to transfer the species from Appendix II to Appendix I. As it is required to do under the text of the Convention, the Secretariat provided its own recommendations to the Parties in relation to the proposal to move the African elephant from Appendix II to Appendix I.[31] These were measured and rather nuanced. The proposal was being strongly supported by a number of conservation and animal welfare NGOs and a rather feverish atmosphere developed amongst the Parties and in the media prior to CoP7 in 1989. Allegations circulating of illegal activities associated within the Secretariat in the form of corruption.[32] NGOs, including the Swiss Bellerive Foundation led by Prince Sadruddin Aga Khan, wrote to UNEP Executive Director Mostafa Tolba complaining about the way that Secretary-General Lapointe and his staff had been conducting themselves in relation to African elephants.

At the close of CoP7 in October 1989, UNEP's representative mounted an extremely spirited defence of the Secretariat and its staff using the strongest of language.[33] The UNEP Executive Director did however request an enquiry into the Secretariat be under-taken by UNEP staff. It uncovered no wrongdoing, but at some stage Tolba decided that he was not going to renew Lapointe's contract as Secretary-General because he had

> lobbied in the media, before the meeting of the Conference of the Parties, to further his [Lapointe's] point of view on proposals put forward on the future status of the African elephant. [Tolba] viewed this as unacceptable and a breach of the Secretary-General's neutrality as a servant of the Parties.[34]

The Parties, in the shape of the Standing Committee, were first informed of the intention of the Executive Director of UNEP to replace the Secretary-General at their meeting in February 1990. The Standing Committee agreed to write to Tolba, recog-nizing that a crisis of confidence in the staff of the Secretariat had been generated at CoP7 in 1989, but saying that they had every confidence in the Secretary-General to conduct business in an honest and effective manner. The Standing Committee began to establish a system for the appraisal of the performance of the Secretariat, and the

Secretary-General in particular. Whilst not questioning the right of the Executive Director of UNEP to hire and fire staff, they did believe that the Standing Committee must have a role in decisions related to the staffing of the position of Secretary-General.

When the post of CITES Secretary-General was advertised on 18 May 1990 with a start date of 1 October that year, the Standing Committee firstly wrote to the Executive Director of UNEP on 1 June 1990 asking him to reconsider and then convened a special meeting to discuss the issue. Despite its unease with the situation, the Standing Committee participated in this recruitment, and Canada, Japan, Malawi, and Switzerland were chosen to represent the Standing Committee on the selection panel. The Selection Panel met twice and, on both occasions, found at least one candidate qualified for the position, but the Executive Director refused to confirm the appointment of the first candidate recommended by the selection panel,[35] heightening the sense of tension. Meantime, Lapointe left the Secretariat on 2 November 1990 even though his contract extended until 31 December.

Later, Lapointe appealed against UNEP's refusal to extend his contract, and a UN internal appeals board concluded that the UNEP Executive Director had acted in an arbitrary and capricious manner and that Lapointe should be reinstated or be paid compensation. By the time this verdict was delivered, a new CITES Secretary-General was in place, and so the latter prevailed.

This crisis was destabilizing for the Convention and its Secretariat but did result in a first, and perhaps overdue, formal Agreement between the CITES Standing Committee and the Executive Director of UNEP over the administration of the Secretariat, signed in 1992.[36] Standing Committee unease about the arrangements for UNEP's hosting of the Secretariat remained, particularly over personnel and financial issues, and a revised Agreement was signed in June 1997.[37] By 2002, the Standing Committee was concerned that UNEP was not complying with the new Agreement and prepared a revised draft. The UNEP Executive Director, however, was unwilling to sign this pending legal advice and the development of a common approach to its hosting of convention secretariats more broadly.

It was not until 2011 that a new Memorandum of Understanding between the UNEP Executive Director and the Standing Committee came into effect.[38] Concerns about UNEP's performance with regard to the provision of the Secretariat continued though, and in 2016 the Standing Committee established a working group to consider the advantages and disadvantages of different administrative hosting models for the Secretariat.[39] This did not lead to any substantive proposals for change. The relationship between the CITES Parties and UNEP over the hosting of the CITES Secretariat has been a constant source of friction.[40]

Following Lapointe's departure, the new CITES Secretary-General, Bulgarian diplomat Izgrev Topkov, started work on 1 July 1991—very soon after Bulgaria itself joined CITES. Although Topkov had been Permanent Representative of Bulgaria to UNEP, he did not have a particular background in environment issues. In May 1993 he supervised the move of the Secretariat from Lausanne to the suburbs of Geneva in Switzerland. The Swiss Depositary Government wished to create an 'environment hub' in Geneva housing a number of different environment organizations and were willing to pay some removal costs and a large proportion of the initial rental charges. The move to what was then called the Geneva Executive Centre (now International

FIGURE 3.3   Maison internationale de l'environnement (International Environment House), Geneva, current headquarters of the Secretariat.

Photo: Geneva Environment Network

Environment House) (Figure 3.3) was supposed to be temporary pending final establishment in the Wilson Palace, an imposing building in the city on the shores of Lake Geneva.[41] In the event, this building was subsequently occupied by the Office of the United Nations High Commissioner for Human Rights, and the CITES Secretariat remained at International Environment House, where it is today.

In more spacious offices and in a more favourable economic climate, staff numbers continued to increase. Topkov and his staff embarked on a drive to increase the public profile of CITES. This included production of two CITES jigsaw puzzles, a board game, and a card game illustrating CITES-listed species with an explanatory note about the Convention. Royalties and percentage of the profits from sales accrued to the Convention. More usually, an agreement was signed with a French vineyard in order to create a special CITES wine, which was reportedly 'a great success'.[42]

In 1993, to celebrate the 20th anniversary of the signing of the Convention, the United Nations Postal Administration, in cooperation with the Secretariat, produced a set of stamps featuring CITES-listed species. New sets of stamps have been produced in most years since then.

In 1994 a glossy magazine, *CITES/C&M International Magazine* (where the C&M stood for conservation and management) was launched,[43] produced by the Secretariat

FIGURE 3.4 *CITES/C&M International Magazine.*

in cooperation with an Argentinian foundation (Figure 3.4). It was presented to Parties at CoP9 later in the year.[44] Produced in English and Spanish, it was initially planned to be a quarterly, but after five issues during its first three years, it ceased publication.

It was replaced in August 1998 by a twice-yearly newsletter, *CITES World*, which Parties requested at CoP10 in 1997.[45] This publication ran for 18 issues until July 2009.

By the end of 1995, the Secretariat had over 20 staff, including two translators (French and Spanish) and a deputy enforcement officer seconded from the UK. Topkov tried to restructure the Secretariat, breaking up the management unit into new units dealing with Enforcement Assistance and Permit Confirmation, Convention Interpretation, Monitoring and Servicing, and a Capacity Building Unit. The revision also included a new post of Deputy Secretary-General from September 1996, which was taken by Australian Jim Armstrong, former Chair of the Plants Committee.

This new structure met resistance from some Secretariat staff even after the reorganization was announced as completed in November 1996. Discord continued and the atmosphere deteriorated with a self-styled 'dissident' group established within the staff. UNEP sent a high-level representative from their headquarters in Nairobi, Kenya, to resolve the situation, but the tensions continued. The Secretary-General took the unusual step of presenting the new work-unit structure to CoP10 in 1997 for endorsement, which was forthcoming, but this did not resolve the discontent and different parts of the Secretariat started to correspond by memorandum from one end of the corridor to the other.

The Executive Director of UNEP convened a panel to enquire into the personnel and management issues, which notably contained no representatives from CITES Parties. The Panel visited the Secretariat in December 1997, and its summary report was excoriating. It was proposed that two staff who were considered at the centre of, and largely the originators of the conflict for several years, should be requested to leave 'in lieu of stronger administrative action'.[46] Secretary-General Topkov was recommended to have his employment extended for one year whilst recruitment of a replacement Secretary-General took place. The Panel also strongly recommended that Jim Armstrong be moved to another post (within the UN system).

The affair rumbled on at the following meeting of the Standing Committee in March 1998 with the situation exacerbated by the regional co-ordinators' roles played by senior Secretariat staff, which resulted in some Parties being briefed privately on the internal situation in the Secretariat rendering frank discussions about the matter between the Parties difficult. The Executive Director of UNEP referred the matter to the UN Office of Internal Oversight Services in New York as an independent auditor, with a request for advice within three weeks. By July 1998 recruitment advertisements for three posts appeared. These were to replace Obdulio Menghi, head of the Scientific Coordination Unit, who took early retirement; Jean-Patrick Le Duc, head of Enforcement Assistance and Permit Confirmation Unit, who was still on secondment from his national government and later returned there; and Secretary-General Topkov, who departed voluntarily in June 1998. The affair though did not end there. *Nature*, one of the world's leading multidisciplinary science journals, published a news item on 9 July quoting a UNEP spokesman who said that two staff had been dismissed 'for improper behaviour in handing out [CITES] permits'.[47] Despite the quotation marks, UNEP said that no officially authorized spokesmen made the statement in question.[48] Nevertheless, this allowed NGOs not favourable to some of the actions of the Secretariat to claim that there was significant malfeasance throughout the staff, including the sale of export permits.[49] Although three staff departed, Armstrong remained in place as Deputy Secretary-General. The whole affair dented the credibility of the Secretariat and took up many hours of time that could have been used to further the Convention's objectives.

The recently retired Deputy Executive Director of UNEP Reuben Olembo was drafted in to serve as Secretary-General ad interim from 5 August 1998 to 31 March 1999. During this time, Willem Wijnstekers, a Dutch official from the European Commission in Brussels, Belgium, was appointed as the new Secretary-General. In contrast to Topkov, he was a CITES enthusiast of long standing, having been the technician responsible for co-ordinating the implementation of the Convention in the European Union (then the European Community). He was also well-known in CITES circles as author of the definitive reference guide to the Convention, *The Evolution of CITES*, first published in 1988,[50] which ran to 11 updated editions with some translated into French, Spanish, and Russian.

The return of some stability was welcomed by Parties, and the subsequent CoP11 in 2000 agreed to a number of new posts including on animal science, documentation, and legal and trade policy. Using these new recruits, Wijnstekers' reorganization of the Secretariat structure, which was not radically different than that proposed by Topkov, was accepted by Secretariat staff without complaint.

As part of the Convention's policy to regulate the trade in African elephants following its transfer to Appendix I, it had been agreed to establish a programme for Monitoring the Illegal Killing of Elephants (MIKE). With external donor funding, a Central Coordinating Unit for the programme was established, attached to the Secretariat but based in Nairobi, Kenya—the first time that Secretariat staff had been located outside Switzerland. Although budgetary constraints continued and the post of Deputy Secretary-General went unfilled after Jim Armstrong left in 2007, a period of stability in the Secretariat followed. In terms of work planning, at CoP14 in 2007, for the first time, the Secretariat proposed a costed programme of work for the upcoming triennium.[51] The purpose, in the context of discussions about the budget, was to increase transparency and try and provide a link between the demands on the Secretariat made by Parties and the resources available to deliver them.[52]

Wijnstekers invested time in building relationships with organizations particularly in the biodiversity sector. He tried to rekindle the historic relationship with IUCN—The World Conservation Union by signing a Memorandum of Understanding (MoU).[53] The Convention on Biological Diversity had agreed to create a Liaison Group of Biodiversity-related Conventions comprised of heads of the secretariats of those conventions, and Wijnstekers was an enthusiastic supporter during its establishment. He signed an MoU with the secretariat of the Convention on the Conservation of Migratory Species of Wild Animals (CMS), which covers many of the same species as CITES, and a joint work programme was established to promote coordination of work on mutually significant key species.

The Secretariat attended the International Conference on Biodiversity, Science, and Governance, held in Paris January 2005,[54] which eventually led to the establishment of the Intergovernmental Science-Policy Platform on Biodiversity and Ecosystem Services (IPBES) to strengthen the science-policy interface for biodiversity. Encouraged by the Secretariat which signed a Memorandum of Cooperation with the secretariat of IPBES in 2017, the Parties engaged with this initiative,[55] principally through the Chairs of the Animals and Plants Committees. IPBES produced a Global Assessment Report on Biodiversity and Ecosystem Services[56] and, after pressure from CITES, an Assessment Report on the Sustainable Use of Wild Species[57] with much food for thought about the approach adopted by the Convention and its place in the wider effort to conserve biodiversity.[58]

Wijnstekers also promoted cooperation on the enforcement of the Convention, creating a separate Enforcement Unit within the Secretariat. He signed an MoU with the secretariat of the Lusaka Agreement on Cooperative Enforcement Operations

Directed at Illegal Trade in Wild Fauna and Flora in Africa, but more significantly for the longer term, he laid the groundwork for the establishment of the International Consortium on Combating Wildlife Crime (ICCWC), a partnership between the CITES Secretariat and INTERPOL, the United Nations Office on Drugs and Crime, the World Bank, and the World Customs Organization. The formal agreement creating ICCWC was signed a few months after his departure from the Secretariat.

In 2005, the Secretariat invited the secretariat of the International Tropical Timber Organization (ITTO) to join it in a capacity-building programme to assist countries to implement the Convention for the growing number of commercial timber species being listed in the CITES Appendices.[59] The ITTO-CITES programme on timber species was launched in 2007 and proved popular with Parties receiving support and with donors.

Wijnstekers' tenure brought stability to the Secretariat after a period of turmoil, and his retirement after CoP15 in 2010 provided an opportunity to recall the huge influence he had on the Convention, both before and after he joined the Secretariat.

His replacement was John Scanlon an Australian lawyer. Although not known to most CITES participants, he had previously been director of the IUCN Environmental Law Centre, keeping up the long tradition of links between the Centre and CITES. He arrived from the senior ranks of UNEP in Nairobi, whom Wijnstekers had kept at arm's length. Suspicions that he might have been sent to bring the CITES Secretariat 'under more UNEP control' were however quickly dispelled. At a time of heightened tension between the CITES Parties and UNEP, he sought to provide clarity over his own position by agreeing a delegation of authority on administrative and financial matters from the Executive Director of UNEP signed in October 2010[60] and an amended form remains in place in 2025.

Scanlon carried out a modest restructuring of the Secretariat, dissolving the specific Capacity-building Unit and charging all professional staff to do this work as part of their day-to-day specialist responsibilities. A Knowledge Management and Outreach Services unit was created to take advantage of new technology, create a proper focus on media and communications issues, and to lead on resource mobilization.

He raised the eyebrows of some for being shown frequently on the CITES website and its social media feeds shaking hands with different VIPs, but his diplomatic and networking skills ensured that the Convention reached new heights in term of its public profile. After inheriting a settled staff team and cultivating cordial relations with all the key players amongst the Parties, he was able to devote time to two key themes: advancing work on enforcement to combat illegal trade and promoting CITES in the wider context of sustainable development.

Scanlon prioritized work on enforcement, and he got off to a good start by convincing the Secretary-General of INTERPOL to sign up to ICCWC in a private on-one-one meeting in Lyon, which was followed by the other executive heads of the proposed consortium. The formal establishment of ICCWC, at a Global Tiger Summit in Saint Peterburg, Russian Federation, in November 2010, proved a huge boost to tackling illegal international trade. ICCWC's mission to strengthen criminal justice systems and provide coordinated support at national, regional, and international level to combat wildlife and forest crime attracted donor support on a scale which quickly dwarfed that of the Secretariat itself. From the start, it was agreed that the chairing of the Senior Experts Group which guides the work of the Consortium should rest with CITES, and the Secretariat housed a small number of staff working exclusively on ICCWC.

In terms of positioning CITES in the wider context of sustainable development, Scanlon set the tone in 2012 when he engineered a specific reference to CITES in the outcome document of the UN Conference on Sustainable Development (Rio+20),[61] which was subsequently adopted by the UN General Assembly:[62]

> We recognize the important role of the Convention on International Trade in Endangered Species of Wild Fauna and Flora, an international agreement that stands at the intersection between trade, the environment and development, promotes the conservation and sustainable use of biodiversity, should contribute to tangible benefits for local people, and ensures that no species entering into international trade is threatened with extinction. We recognize the economic, social and environmental impacts of illicit trafficking in wildlife, where firm and strengthened action needs to be taken on both the supply and demand sides. In this regard, we emphasize the importance of effective international cooperation among relevant multilateral environmental agreements and international organizations. We further stress the importance of basing the listing of species on agreed criteria.

In 2015, two targets were inserted into the UN 2030 Agenda for Sustainable Development which had direct relevance to the mandate of CITES[63]:

> Take urgent action to end poaching and trafficking of protected species of flora and fauna and address both demand and supply of illegal wildlife products.
>
> Enhance global support for efforts to combat poaching and trafficking of protected species, including by increasing the capacity of local communities to pursue sustainable livelihood opportunities.

Further references to CITES were made in a UN General Assembly Resolutions on illicit trafficking in wildlife.[64] At CoP16 in 2013, Scanlon persuaded Thailand to spearhead a plan to declare 3 March (the date, in 1973, on which CITES was adopted) as World Wildlife Day.[65] Parties agreed, and in doing so, recalled that the Plenipotentiary Conference had been referred to at the time as the 'World Wildlife Conference'.[66] Subsequently the UN General Assembly proclaimed 3 March as such, requesting the Secretariat to facilitate its implementation. More controversially, Scanlon's nose for publicity led him to attend the public destruction of seized elephant ivory or rhino horn in China and Hong Kong, the Czech Republic, Kenya, Sri Lanka and elsewhere. Whilst he was careful to stress that such destruction was a matter for each country to determine, it was hard not to see his presence as an endorsement of the action.

What was not in doubt was the publicity for CITES raised by these events, nor the opportunity they afforded to cultivate high-level contacts in the Parties concerned. Notable amongst these was China, which in view of its increased economic power and traditional use of wildlife for food and medicine, was emerging as the main destination for trade in many CITES-listed species. Recognizing this fact, Scanlon invested considerable effort to develop close relations with China, visiting the country eight times during his eight-year tenure. He awarded their National Inter-Agency CITES Enforcement Collaboration Group a Certificate of Commendation[67] and attended a National Retreat for its Management Authority staff—the largest CITES Management Authority in the world. Whilst China continues to be a major market for international wildlife trade (legal and illegal), its changes of policy on issues such as ivory trade may be a reflection of Scanlon's efforts.

From an administrative perspective, the Secretariat had been fairly independent of UNEP, but this changed in 2015, with the arrival of a UN-wide resource planning system named 'Umoja' (meaning 'unity' in Kiswahili language). It centralized many administrative processes across the UN, including finance/accounting, procurement, and human resources. Designed to 'streamline fragmented administrative processes to allow Managers and Staff to focus on [their] mandates rather than lengthy approval processes',[68] it had quite the reverse effect. With its ponderous computer interface, it required many more steps to undertake even the most basic task. Whilst its cost and benefit were being criticized more widely,[69] within the Secretariat with its rapidly growing workload, and being used to a more autonomous way of working, the effect on staff was particularly acute.[70]

Scanlon's energy and leadership were widely admired,[71] and so his decision to leave the Secretariat in April 2018 took many by surprise. He left a legacy of action and activity that will be hard to equal.

After an interregnum, he was replaced in December 2018 by Ivonne Higuero, a US-educated Panamanian economist. She became the first woman to lead the Secretariat—after 43 years. She was also the first career UN civil servant to hold the post, and she embraced many of the bureaucratic UN processes that previous Secretary-Generals had kept at arm's length. This tendency has also been noted outside the Secretariat and has been dubbed by some CITES meeting delegates as 'the UN-ification of CITES'.[72]

Higuero suffered some misfortune with her first two meetings of the CoP. In her first week of work, she went to Sri Lanka to galvanize flagging Government efforts to organize CoP18 there in 2019. However, just when things were looking positive, a wave of bombings hit the Sri Lankan capital Colombo in April 2019. Despite Sri Lanka's efforts to reassure them, Parties were not convinced that the CoP could be held in good conditions, and it was instead moved to Geneva at short notice with the financial support of Switzerland. The following CoP was due to be held in Costa Rica, but they withdrew their offer due to the economic impact of the COVID-19 pandemic. Fortunately for CITES, Higuero was able to use her Panamanian connections, and CoP19 was held in Panama City in November 2022.

Like many other aspects of life, the work of the Secretariat was severely disrupted by the COVID-19 pandemic. Staff worked mostly from home for a little over two years. Meetings of the CITES intersessional committees scheduled for 2020 had to be postponed, but after extensive planning, they were organized online in 2021. This proved remarkably successful, particularly in terms of the number of Parties and observers, and of delegates, participating. The Secretariat also successfully delivered a number of other technical and training workshops online in 2020 and 2021. In the early days of the pandemic, international trade in wildlife was talked about as a possible contributory factor to the spread of the SARS-CoV-2 virus, and there were rather panicked calls for CITES to do something about it. With considerable wisdom and foresight, the Secretariat issued a statement on 17 March 2020 recalling that matters regarding zoonotic diseases were outside CITES's current mandate and the Secretariat's competence.[73]

Later, a more measured approach was agreed by Parties to review these linkages.[74] In what may have been seen as an act of premonition, the Secretariat had established a Cooperation Agreement with the World Organisation for Animal Health in 2015, being the primary intergovernmental organisation in the field of animal health, which recognised the risk of zoonotic diseases. At the direction of the Parties, this was revised into a Memorandum of Understanding in 2024 in light of the experiences gained during the COVID-19 pandemic. The Secretariat also undertook a survey of the impact of the pandemic on the working practices of national CITES authorities and prepared a detailed review of the implications of the COVID-19 pandemic for the implementation of the Convention.[75] However, there was little appetite amongst the Parties for exploring the option of online meetings, despite their likely benefits in terms of outreach, cost-effectiveness, and reducing the Convention's environmental impacts.

Since its present core size stabilized in the 1990s, the Secretariat has been divided into three to five teams with a relatively flat organizational structure which contributed to a 'family' atmosphere. However, UN-wide mandatory protocols on annual staff reporting became steadily more complicated, and the administrative burden on staff managers grew. Higuero decided a more pyramidal organizational structure was needed. This and the other bureaucratic hoops that Secretariat staff were increasingly obliged to jump through, particularly resulting from the effects of Umoja, resulted in less time being available for dialogue between in-house teams and for peer review of work being done. The pressure was telling, and there was an unprecedented wave of voluntary staff departures from the Secretariat.

At CoP19 in 2022 a remarkable 367 decisions were agreed, many directed to the Secretariat. This increase in activities is also reflected at meetings of the CoP and intersessional committees where the agendas have become huge. As the Secretariat writes most of the documents considered in these forums, this adds considerably to the its burden (Figure 3.5).

3.3   BUDGET

The cost of the Secretariat is paid for from the CITES Trust Fund. Three quarters of the total Trust Fund budget is for the provision of the Secretariat, most of the rest is spent on organizing meetings and given to UNEP as an overhead. The budget for each triennium is agreed at each CoP and as years in which the CoP is held cost more than other years, an average annual figure for the triennium is the best way to look at changes in the budget (Figure 3.6). The first, which started from 1 July 1975, was 322,244 USD per year.

The chronic underfunding of the Secretariat came to a head at CoP6 in 1987. Faced with the prospect of some drastic cutbacks,[76] the CoP agreed to an increase of over 100% for 1988–1989 compared with 1986–1987.[77] Further increases followed at the subsequent two CoPs. These also accommodated a UNEP decision to upgrade the level of several senior posts after a period of staff turbulence. Whilst the Trust Fund budget has been increased at almost every meeting of the CoP, for context, it remains about a tenth of the production cost of a single jet fighter plane.

FIGURE 3.5   CITES Secretary-Generals: Clockwise from the top: Ivonne Higuero, John Scanlon, Willem Wijnstekers, Izgrev Topkov, Eugène Lapointe, and Peter Sand.

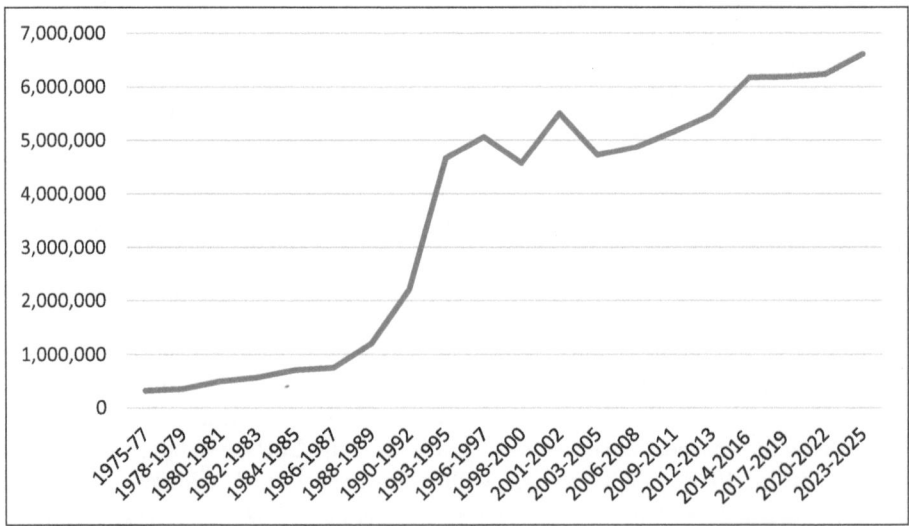

FIGURE 3.6   Averaged annual budget of the CITES trust fund (or funding for the Secretariat from UNEP before 1980).

3.4 PARTNERSHIPS

The Secretariat's resources have never been enough to achieve all the results requested of it by Parties. Thus, it has partnered with many organizations in order to achieve these tasks. Three organizations have been particularly important in this regard. IUCN played a profound role in the creation of the Secretariat, and until 1984 its staff acted as the Secretariat, under the responsibility of UNEP. After UNEP took over the full responsibility for the Secretariat in 1984, the relationship with IUCN weakened a little until a new Memorandum of Understanding was signed in 1999 in which the Secretariat nominated IUCN as a 'major technical advisor to the Convention'. IUCN's particular inputs came mainly through its Species Survival Commission and included its analyses of proposals to amend the Appendices, contribution to the science of non-detriment findings, Review of Significant Trade, and other ad hoc activities under contract. On the ground, IUCN played a major part in developing the Monitoring the Illegal Killing of Elephants programme. Whilst IUCN remains an important partner in CITES implementation, in recent years it has turned its focus more on the Convention on Biological Diversity as a more holistic treaty more likely to positively affect global biodiversity on a significant scale.

TRAFFIC began as a Specialist Group of IUCN's Survival Service Commission and was the first consultant to undertake a comprehensive analysis of annual trade reports from Parties.[78] In partnership with WWF, it was developed into a separate organization and was a real pioneer in on-the-ground research into wildlife trade with its original remit to monitor international wildlife trade and submit its findings to inform decisions made by CITES Parties. Its publication TRAFFIC Bulletin, begun in 1979, was essential reading for early CITES practitioners and was sent out by the Secretariat in Notifications to the Parties (Figure 3.7). The Secretariat used TRAFFIC as a consultant for a wide range of activities when there were few other organizations with expertise in wildlife trade. In 1997, the CoP recognized the Bad Ivory Database System (later renamed the Elephant Trade Information System), established by TRAFFIC in 1992, as the reference for compiling law enforcement data on seizures and confiscations of ivory and other elephant specimens, and it became a significant tool for monitoring illegal trade in elephant ivory. A Memorandum of Understanding between the Secretariat and TRAFFIC in 1999 focussed on capacity-building. With its global network of offices, TRAFFIC was better placed to keep closer contact with Parties than the Secretariat. Staff of TRAFFIC regularly moved on to fulfil roles in national CITES offices and several, such as Jonathan Barzdo, Tom De Meulenaer, and Steve Nash, went on to occupy senior roles in the Secretariat. The ties between the Secretariat and TRAFFIC have weakened in recent years as TRAFFIC has broadened its interests, and the Secretariat sought to be more independent of an increasingly crowded field of NGOs.

Another key relationship that began with IUCN is with what is now known as the UN Environment Programme World Conservation Monitoring Centre (UNEP-WCMC). UNEP-WCMC started as the Species Conservation Monitoring Unit of IUCN in 1979,[79] and its wildlife trade monitoring section took over from TRAFFIC in analyzing Parties' annual trade reports.[80] The IUCN's Species Conservation Monitoring Unit has undergone a variety of organizational changes over the years and in 2000 it became part of UNEP in conjunction with a UK charitable institution.

FIGURE 3.7   CITES Secretariat staff at CoP18 in 2019.

Photo by IISD/ENB I Kiara Worth

UNEP-WCMC's early use of computer systems led to the development of what is now the CITES Trade Database,[81] which houses all annual trade reports submitted by Parties on behalf of the Secretariat. In recent years, UNEP-WCMC has also been commissioned to undertake other work by the Secretariat, notably the data management aspects of the Review of significant trade and the Review of trade in animal specimens reported as produced in captivity compliance processes. The impartiality and expertise of its staff is respected by Parties, and it also works as a data manager for other biodiversity conventions, which enables it to suggest and promote synergy between them and CITES.

NOTES

1 United Nations Environment Programme (2024) *United Nations Environment Assembly.* Sixth session Nairobi, Kenya 26 February–1 March 2024. Document UNEP/EA.6/INF/6.

2 United Nations General Assembly Resolution 2997. *Institutional and Financial Arrangements Lor International Environmental Cooperation.* https://documents.un.org/doc/resolution/gen/nr0/270/27/pdf/nr027027.pdf (accessed 25.02.25).

3 Klimke, V. (2015) *A Sustainable Life. Wolfgang E. Burhenne and the Development of Environmental Law.* Privately Published.

4 Anon (1976) *No. 14537 Convention on International Trade in Endangered Species of Wild Fauna and Flora (with Appendices and Final Act of 2 March 1973). Opened for Signature at Washington on 3 March 1973.* United Nations-Treaty Series Vol. 993:

243–417. https://treaties.un.org/doc/publication/unts/volume%20993/volume-993-i-14537-english.pdf (accessed 25.02.25).

5 United Nations Environment Programme. *Report of the Governing Council on the Work of Its First Session 12–22 June 1973*. United Nations Governing Council twenty-eight session. Supplement No. 25 (A/9025), New York.https://documents.un.org/doc/undoc/gen/n73/178/33/pdf/n7317833.pdf (accessed 25.02.25).

6 Doc. 1.8.

7 Resolution Conf. 1.8 Annex.

8 United Nations Environment Programme. *Report of the Governing Council on the Work of Its Fifth Session 9–25 May 1977*. United Nations Governing Council Thirty-Second Session. Supplement No. 25 (A/32/25), New York.https://documents.un.org/doc/undoc/gen/n77/164/44/pdf/n7716444.pdf (accessed 25.02.25).

9 *CITES Secretariat 2nd Annual Report*. https://cites.org/sites/default/files/document/E-Annual%20Report%201977.pdf (accessed 25.02.25).

10 Anon (2015) *Peter H. SAND—Curriculum Vitae*. Ludwig-Maximilians-Universität München, Munich, Germany. https://lmu-munich.academia.edu/PeterHSand/CurriculumVitae (accessed 25.02.25).

11 SC66 Inf. 1.

12 Doc. 2.5.

13 Resolution Conf. 2.1.

14 Doc. 2.10.

15 Doc.2.6 Annex 2.

16 Notification to the Parties N° 139 of 5 June 1980.

17 *Canadian Wildlife Service Annual Review 1982–1983*. https://publications.gc.ca/collections/collection_2023/eccc/CW70-24-1983-eng.pdf (accessed 25.02.25).

18 Doc. 3.5.

19 Holgate, M. (1999) *The Green Web: A Union for World Conservation*. Earthscan Publications Ltd, London.

20 SC10 Summary Record.

21 Doc. 5.8.

22 Doc. 6.6 Annex 2.

23 Doc. 6.6.

24 Resolution Conf. 5.12.

25 Gup, T. (1989) Elephants: Trail of Shame. *Time Magazine*, 16 October. https://time.com/archive/6703598/elephants-trail-of-shame (accessed 25.02.25).

26 Doc. 7. 21.

27 Doc. 7.11.

28 Doc. 8.6.

29 Plen. 7.2.

30 Doc. 7.7.1 Annex.

31 Doc. 7.43 Annex 3.

32 United States District Court for the District of Columbia. Eugene Lapointe, Plaintiff, Craig van note, et al. Defendants. Civil action no. 03–2128 (rbw). https://ecf.dcd.uscourts.gov/cgi-bin/show_public_doc?2003cv2128-74 (accessed 25.02.25).

33 Plen. 7.8 Annex.

34 United Nations Joint Appeal Board (1993) *Case of Eugene Lapointe. Report to the Secretary-General*. Report 991. Case N° 92–66. www.iwmc.org/wp-content/uploads/2021/06/UN-Joint-Appeals-Board-EL-Case.pdf (accessed 25.02.25).

35 Doc. 8.5.

36 CITES Secretariat (1992) *Agreement between the CITES Standing Committee and the Executive Director of UNEP.* https://cites.org/sites/default/files/eng/cop/08/E-Agreement-CITES_UNEP.pdf.

37 SC47 Doc. 7.

38 CITES Secretariat (2011) *Memorandum of Understanding between the Standing Committee of the Conference of the Parties to the Convention on International Trade in Endangered Species of Wild Fauna and Flora and the Executive Director of the United Nations Environment Programme Concerning Secretariat Services to and Support of the Convention.* https://cites.org/sites/default/files/common/disc/sec/CITES-UNEP_0.pdf (accessed 25.02.25).

39 SC67 Summary Record.

40 SC66 Inf.1.

41 SC29 Summary Report.

42 Doc. 10.8 (Rev.).

43 Notification to the Parties No 813 of 21 July 1994.

44 Plen. 9.4 (Rev. 2).

45 Dec. 10.104.

46 UNEP (1998) *Summary Report of the Panel on the Enquiry into Personnel and Management Issues in the CITES Secretariat* (Geneva, 8–13 December 1997). 16 January 1998.

47 Masood, E. (1998) CITES Chief Removed in Scandal Over Trade in Banned Species. *Nature*, 394: 112.

48 Barabanov, A. (1998, September 1) *Chief of Administrative Services.* UNEP. in litt.

49 Van Note, C. (2003) Victories at CITES. *Earth Island Journal.* Spring. www.earthisland.org/journal/index.php/magazine/entry/victories_at_cites (accessed 25.02.25).

50 Wijnstekers, W. (1988) *The Evolution of CITES.* Secretariat of the Convention on International Trade of Endangered Species of Wild Fauna and Flora, Lausanne, Switzerland.

51 CoP14 Doc. 7.3 (Rev. 1).

52 SC54 Doc. 6.2.

53 CITES Secretariat (1999) *Memorandum of Understanding between the Secretariat of the Convention on International Trade in Endangered Species of Wild Fauna and Flora and IUCN-the World Conservation Union July 1999.* https://cites.org/sites/default/files/eng/disc/coop/CITES-IUCN.pdf (accessed 25.02.25).

54 Le Duc, J.-P. (Ed.) (2005) *Proceedings of the International Conference on Biodiversity, Science and Governance.* Muséum National d'Histoire Naturelle, Paris, France.

55 Decisions 15.12–15.14, 16.13–16.15 and Resolution Conf. 18.4.

56 IPBES (2019) *Global Assessment Report on Biodiversity and Ecosystem Services of the Intergovernmental Science-Policy Platform on Biodiversity and Ecosystem Services* (pp. 1–1082). Brondízio, E.S., Settele, J., Díaz, S. and Ngo, H.T. (Eds.). IPBES Secretariat, Bonn, Germany. https://doi.org/10.5281/zenodo.3831673.

57 IPBES (2022) *Thematic Assessment Report on the Sustainable Use of Wild Species of the Intergovernmental Science-Policy Platform on Biodiversity and Ecosystem Services.* Fromentin, J.M., Emery, M.R., Donaldson, J., Danner, M.C., Hallosserie, A. and Kieling, D. (Eds.). IPBES Secretariat, Bonn, Germany. https://doi.org/10.5281/zenodo.6448567.

58 Decs. 19.28–19.29.

59 Johnson, S., Sosa Schmidt, M. and Carrillo, R. (2016) *ITTO and CITES: An Enduring Partnership.* ITTO Tropical Forest Update. Vol. 25 No. 1.

60 United Nations Environment Programme (2010) *Delegation of Authority on Administrative and Financial Matters From: Achim Steiner, Executive Director of the*

United Nations Environment Programme (UNEP) to: John Scanlon, Secretary-General of the Convention on International Trade in Endangered Species of Wild Fauna and Flora (CITES) https://cites.org/sites/default/files/common/disc/sec/delegation_authority_0.pdf (accessed 25.02.25).

61 Anon (2012) *Report of the United Nations Conference on Sustainable Development Rio de Janeiro, Brazil 20–22 June 2012.* United Nations, New York. https://documents.un.org/doc/undoc/gen/n12/461/64/pdf/n1246164.pdf (accessed 25.02.25).

62 United Nations (2012) *Resolution Adopted by the General Assembly on 27 July 2012. 66/288. The Future We Want.* United Nations General Assembly Sixty-Sixth Session A/RES/66/288.   https://documents.un.org/doc/undoc/gen/n11/476/10/pdf/n1147610.pdf (accessed 25.02.25).

63 United Nations (2015) *Resolution Adopted by the General Assembly on 25 September 2015. 70/1. Transforming Our World: The 2030 Agenda for Sustainable Development.* United Nations General Assembly Seventieth Session. A A/RES/70/1. https://documents.un.org/doc/undoc/gen/n15/291/89/pdf/n1529189.pdf (accessed 25.02.25).

64 United Nations (2021) *Resolution Adopted by the General Assembly 23 July 2021. 75/311. Tackling Illicit Trafficking in Wildlife United Nations General Assembly Seventy-Fifth Session.*   A/RES/75/311.   https://documents.un.org/doc/undoc/gen/n21/205/05/pdf/n2120505.pdf (accessed 25.02.25).

65 CoP16 Doc. 24 (Rev. 1).

66 Resolution Conf. 16.1.

67 CITES Secretariat (2012) *New National CITES Enforcement Coordinating Body Shows Positive Results.* https://cites.org/eng/news/pr/2012/20120509_certificate_cn.php (accessed 25.02.25).

68 Anon (2016) *UMOJA Introduction for Staff.* https://umoja.un.org/sites/umoja.un.org/files/intro_to_staff__1.pdf (accessed 25.02.25).

69 Lynch, C. (2016) *At the United Nations, Umoja Translates as Bureaucratic Chaos.* Foreign Policy. No. 218, May/June 2016. https://foreignpolicy.com/2016/05/06/at-the-united-nations-umoja-translates-as-bureaucratic-chaos/ (accessed 25.02.25).

70 SC66 Doc. 9.2.

71 CITES Secretariat (2018) *Farewell Messages for John E. Scanlon.* https://cites.org/eng/sg/farewell-messages/john-e-scanlon (accessed 25.02.25).

72 Harris, K., Beintema, N. and Rosen, T. (Eds.) (2025) Summary of the 78th Meeting of the CITES Standing Committee: 3–8 February 2025. *Earth Negotiations Bulletin*, 21 (114).

73 CITES Secretariat (2020) *CITES Secretariat's Statement in Relation to COVID-19.* https://cites.org/eng/CITES_Secretariat_statement_in_relation_to_COVID19 (accessed 25.02.25).

74 Notification to the Parties No. 2021/031 of 8 April 2021.

75 CoP19 Doc. 24.

76 Plen. 6.4 (Rev.) Annex.

77 SC54 Doc. 6.2 Annex.

78 Doc. 3. 21.

79 Holgate, M. (1999) *The Green Web: A Union for World Conservation.* Earthscan Publications Ltd, London.

80 Doc. 4.18.

81 CITES Secretariat (2024) *Full CITES Trade Database Download Available* (version 2024.1). https://trade.cites.org/ (accessed 25.02.25).

4 CITES Species

4.1 TAXONOMY AND NOMENCLATURE

Biodiversity can be described at the genetic, species, or ecosystem level, but the species level is the one that resonates most with the general public and is embedded in human culture in all parts of the world. Species are the currency of CITES. The most common description of the term 'species' is of a group of similar individuals that live in the same geographical area and can interbreed to produce fertile offspring. However, whilst species are a fundamental taxonomic unit of biological classification, there are a number of different concepts to define a species,[1] and even then, there are sometimes differences of opinion about which species any given individual specimen may belong to. For a convention that requires mutual recognition of documents that permit international trade in specimens of a species, it was immediately clear to CITES Parties that some common agreement was required about what species would be recognized as valid and which specimens belong to that species. At the first meeting of the CoP in 1976, the Secretariat raised the problem posed for customs officials by referring to the synonyms and differing classification systems being used by Parties. The Secretariat favoured use of the original Latin species names that were the most widely accepted, but most Parties preferred using the most up-to-date scientific names.[2] Finding a balance between these two considerations has been a constant challenge ever since, accentuated by the speed of scientific discovery which has led to a new understanding of the relationships between different species and many changes in the names of species listed in the Appendices. Not only was there a need to harmonize the understanding of the individual species names included in the Convention's Appendices, but there was also a need to agree which species were included and recognized under higher taxonomic listings which had been included in the Appendices. Primates spp. (apes and monkeys) and Psittaciformes spp. (parrots and cockatoos), amongst animal taxa, and Orchidaceae spp. (orchids) and Cactaceae spp. (cacti) amongst the plants are the largest higher taxonomic listings which this concern applies.

The Ad Hoc Committee on Standardized Taxonomy, later to be named the Nomenclature Committee, was established to lead the work of developing a standard list of species whose trade is regulated under the Convention, based, insofar as possible, on rulings of the International Commission on Zoological Nomenclature and International Code for Botanical Nomenclature. A provisional Standardized Taxonomy of Vertebrate Species was produced for comment at CoP2 in 1979,[3] but it was not until CoP4 in 1983 that the first Resolution on standard nomenclature was adopted.[4] It used the names found in a published reference on mammal species and, as an interim aid, a similar reference for bird names. At the same time, the Parties agreed that some changes be made to the names of species in the Appendices to reflect current names of species included therein. In doing so they agreed a cardinal rule that has been applied since: no change to reflect currently understood species

DOI: 10.1201/9781003542278-4

nomenclature may alter the original intent of the proposal to list those animals and plants concerned in the Appendices. If it does, then a fresh proposal to list the taxon in the Appendices must be made.

In 1979 the UK's Scientific Authority for animals began publishing a series of four books which included the adopted names of the animal species listed in the Appendices, the common names in English, and the range States in which they occurred in the wild.[5] Most of these works were commissioned from the IUCN Conservation Monitoring Centre. From 1992, the UK's Scientific Authority for plants produced a series of checklists for various plant taxa.[6] These efforts contributed to a Checklist of CITES Species, first published by the Secretariat in 1996[7] and then regularly updated in printed form until 2013 when it was replaced by a searchable online version.[8] Mention should be made of an earlier checklist of CITES animals and plants, two editions of which were published by the Secretariat in 1990 and 1992.[9] These included English vernacular names and some scientific synonyms for listed species, but no details of the countries in which the species occur.

The Nomenclature Committee, and from 2007 the Animals and Plants Committees, have continued to revise and enlarge the scope of the lists of standard nomenclatural references and propose them for adoption by the Conference of the Parties. Virtually all species listed in the Appendices are now covered by such references. From 1994 onwards, Parties have recommended the use of the 'official' names in the adopted checklists when issuing permits and certificates.[10] In recent years, developments in molecular science have significantly altered our understanding of the relationships between species and their delimitation. These developments have occurred rapidly, with information often published not in book form, but in online references updated in real time. This presents challenges for CITES Parties which only meet in plenary session to consider the possible adoption of such changes every two to three years. The current speed of scientific development may be too rapid for the administrative and enforcement authorities charged with putting the Convention into effect to adapt. Although Parties have provided no guidance, the science committees of the Convention have to balance scientific accuracy with the practicality of implementation when making their recommendations for the adoption of standard nomenclatural references.

4.2   DECIDING WHICH SPECIES' TRADE SHOULD BE REGULATED—THE LISTING CRITERIA

The first list of species to be included in Appendices I and II was drafted by what is now the IUCN Species Survival Commission and was based on what is now known as the IUCN Red List of Threatened Species. At the Plenipotentiary Conference, the voice of the future Parties to the Convention on this point was added by two drafting committees which advised on animals and on plants to be listed in the Conventions Appendices—the drafting committees were chaired by Prof. Jorge Ibarra (Guatemala) and Mr. William Hartley (Australia) respectively.[11] The result

TABLE 4.1

Comparison of the Composition of Appendices I and II between the Original Version in 1975 and that post-CoP19 in 2022

	Fauna			Flora		
	App. I	App. II	App. III	App. I	App. II	App. III
Original Appendices (1975)	489	515	16	246	32,126	6
Post-CoP19 (2024)	734	5,482	394	411	33,764	135
Percentage increase	50%	964%	2362%	67%	5%	2150%

Source: UN Environment Programme World Conservation Monitoring Centre (UNEP-WCMC) with data
extracted from the Checklist of CITES Species

was a total of 33,376 species listed in Appendices I and II. Comparisons with today's figure is difficult, in part due to changes in the understanding of how many species are included in higher taxonomic listings, but the comparable figure after CoP19 in 2022 was 40,391, representing an increase of around 21% (Table 4.1). The increase in the number of species listed in each Appendix was not however spread evenly. At intervening meetings of the CoP, a substantial number of animal species have been added to Appendix II and the number of animal species listed in Appendix I has also gone up by over 50%.

Whilst the species to be listed in Appendices I and II are decided by the Parties collectively, those listed in Appendix III are determined by individual Parties. These Parties seek the cooperation of others in order to control trade in species whose 'exploitation' is prevented or restricted by their national laws. After the entry into effect of the Convention, Canada was the first Party to request inclusion of species in Appendix III in 1975,[12] but for long periods during the first 50 years of the Convention, Appendix III has been relatively infrequently used. In the early years of the Convention, India was the Party who made most use of the possibility to list species in Appendix III.[13] Whilst originally designed to support Parties' national laws to regulate trade in species that may not meet the criteria for inclusion in Appendices I or II of CITES, Appendix III has latterly been used as a prelude to such inclusions. The first and perhaps the most notable example is the African elephant *Loxodonta africana* which was included in Appendix III at the request of Ghana in February 1976. The Giant panda *Ailuropoda melanoleuca* was first included in Appendix III at the request of China, before its inclusion in Appendix I in March 1984. In the case of some tree species and sharks, such listings appear to have been used more strategically to prepare the ground for future proposals to include the species in Appendix II. Notable examples include the Big-leaf mahogany *Swietenia macrophylla* in 1995[14] and Basking shark *Cetorhinus maximus* in 2000.[15] The use of Appendix III to support national regulation of trade has not always been successful. South Africa requested the listing of South African abalone *Haliotis midae* in Appendix III in 2007,[16] but this did not reduce illegal trade[17] and it was withdrawn from Appendix III in 2010.[18]

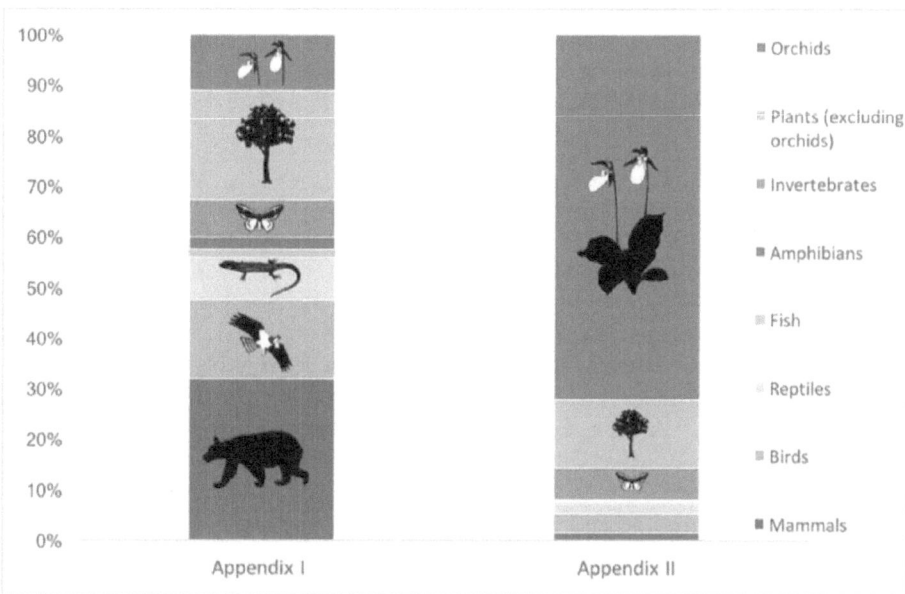

FIGURE 4.1 Composition of Appendices I and II by large taxonomic groups.
Graphic courtesy of Amy Hinsley[19]

From the start, the total number of species covered by the provisions of the Convention was greatly inflated by the inclusion of all members of the orchid family (Orchidaceae spp.—nearly 30,000 species) and cactus family (Cactaceae spp.—around 1,500 species) in Appendix II (Figure 4.1).

When CoP1 began to consider the many proposals that had been submitted to amend the initial allocation of species to the Appendices, it was quickly recognized that the brief fundamental principles in Article II of the Convention, concerning what level of threat merited inclusion in Appendices I or II, did not provide objective criteria upon which to judge proposals. The USA took the initiative and proposed more detailed criteria for determining whether a species qualified for inclusion in Appendix I or II. These criteria were divided into two parts: biological criteria, reflecting the extinction threat to which the species was exposed, and trade criteria reflecting the degree to which the species was subject to international trade.[20] These guidelines to further help assess and judge proposals, which became known as the Berne Criteria were the very first resolution adopted by a CITES CoP[21]. At CoP2 in 1979, these criteria were later complemented by a recommended standard format in which proposals to amend the Appendices should be submitted to a CoP.[22] Concerns about applying the same criteria for judging proposals to reduce CITES protection for species ('downlisting') as to increase ('uplisting') led to the adoption of a second resolution at the same CoP which resolved that deletion from the Appendices or transfer from Appendix I to Appendix II, should require positive scientific evidence that the plant or animal can withstand the exploitation resulting from the removal of protection. As the resolution explains, if an error is made by the Conference by unnecessarily placing a plant or animal on an appendix, the result is the imposition of a documentation requirement. If, however, it errs in prematurely removing a plant or animal from protection, or lowering the level of protection afforded, the result can be the permanent loss of the resource. If it errs it should be therefore toward protection of the resource.

This might be considered an early attempt to apply the 'precautionary principle'. The dichotomy between the criteria for uplisting and downlisting created an imbalance that was not to the liking of all Parties, and the need to revise the Berne Criteria became apparent. In the early years, more radical suggestions for listing species in the Appendices were also explored. Australia proposed that instead of increasing the size of the Appendices by adding extra species, a concept of 'reverse' or 'clean' listing could be considered, where all trade was prohibited except in certain species agreed by the CoP.[23] After a review of the issue, no other Parties were in favour of this approach as it was likely to lead to an increased workload regulating trade in many commoner species.[24]

The Berne Criteria did add helpful interpretation guidance to the basic requirements of the Convention, but in certain circumstances they were considered too rigid. Consequently, under a separate resolution, a process for conditional downlisting was developed. This was first discussed at CoP2 in 1979 where a working group acknowledged that there may be some populations of Appendix I species that can withstand a certain level of commercial exploitation. In particular this applied where specimens of the species were being reared or raised in captivity from eggs or young animals taken from the wild—a practice that the working group referred to as 'ranching'.[25] A resolution about downlisting for ranching was subsequently agreed at CoP3 in 1981. This allowed, subject to certain conditions, consideration of the downlisting to Appendix II of populations of species included in Appendix I which were no longer endangered and could benefit from ranching (the rearing in a controlled environment of specimens taken from the wild).[26] Such downlistings became frequent for crocodilian species between CoP4 in 1983 and CoP10 in 1997. More controversial were proposals relating to ranching of green sea turtle *Chelonia mydas*. France, Suriname, and the UK all put forward amendment proposals to reduce protection for populations of this species (France did so three times) under the ranching provisions, but each time the proposals were either rejected or withdrawn. Amongst the argument made against these proposals were the inadequate identification marking of products planned to be traded from these populations and the possible stimulation of trade in green sea turtle (*C. mydas*) products from parts of its range not covered by the downlisting. No doubt the rather totemic status of sea turtles in conservation circles also mitigated against acceptance of any of these proposals. After initially being in favour of sea turtle ranching if the resulting products replaced wild turtle products in existing local markets, IUCN later decided not to support any attempts to ranch sea turtles unless this would lead to the cessation of commercial exploitation of wild sea turtles.[27] Although the provision to propose downlisting for the purpose of ranching continues today,[28] since CoP10 in 1997 only four such proposals have been made. For reasons of breeding biology, the practice of ranching is best suited to sea turtles and crocodilians. Proposals related to the former did not find favour amongst Parties, and for crocodilians a new process was created to allow downlistings of certain populations.

This new process was a temporary mechanism which was agreed at CoP5 in 1985 to allow species inappropriately listed in Appendix I to be transferred to Appendix II. It was developed because some of the species originally listed when the Convention came into force had never been assessed against the Berne Criteria, and some appeared to have recovered significantly since their inclusion despite baseline data on population status not being available when they were included in Appendix I. This mechanism allowed downlisting from Appendix I to Appendix II subject to various conditions, the principal one being the establishment of an export quota system deemed by the CoP to be sufficiently safe so as not to endanger the survival of the species in the wild.[29] This quota system was used extensively between CoP6 in 1987 and CoP9 in 1992 for downlisting, mostly for crocodilian species. Five proposals concerning leopards *Panthera pardus* were also proposed under this mechanism. These were rejected, although a number of Parties were already able to trade in leopard hunting trophies and skins for personal use whilst the species remained in Appendix I under another resolution.[30]

Notwithstanding these piecemeal changes, some dissatisfaction continued with the Berne criteria. At CoP8 in 1992 a number of countries from the Southern African Development Community (Botswana, Malawi, Namibia, Zambia, and Zimbabwe) submitted that trade in specimens of endangered species should not automatically be considered a threat, rather its impact should be judged empirically. They proposed a new set of listing criteria to reflect this thinking and which they claimed were more quantitative and objective, being based on the emerging new threat criteria being developed by IUCN for its Red List of Threatened Species. The proposed new criteria would be symmetrical, that is to say the same criteria would apply for increasing protection as for decreasing it, unlike the existing criteria at the time which were biased in favour of uplisting. They also envisaged that the Appendices should not be amended without the support of the range States in which the species occurred.[31]

Their proposal was not agreed at CoP8, despite recognition that some species may still not be appropriately listed in the Appendices. Many may not be threatened by commercial trade, and yet attempts to downlist or delete such species had failed. As the Berne criteria were deemed an inadequate basis for amending the appendices, a process was agreed for replacing them.[32] IUCN was commissioned to prepare draft new criteria. After extensive discussions in the Standing Committee and a major workshop in 1993 involving the members and alternate members of the Plants and Animals Committees and members of the Standing Committee, a draft resolution on new listing criteria was submitted to CoP9 in 1994 and agreed by Parties.[33] In line with some of the hopes of the initiators in 1992, the new criteria were much more quantitative and objective than their predecessors and for the most part the same criteria applied for increasing or decreasing the degree of protection afforded to the species by the Convention. However, this was conditioned by the application of a precautionary principle to any change in the listing of a species such that 'scientific uncertainty should not be used as a reason for failing to act in the best interest of the conservation of the species'. Not for the first time in CITES, a form of words was found that could bring comfort to all strands of opinion but arguably would not solve the fundamental differences of view between those Parties broadly in favour of using commercial trade as a conservation tool and those broadly against such an approach.

The apparent success of adopting much more objective listing criteria at CoP9 in 1994 was short-lived. Parties unsuccessfully proposing changes to the Appendices at meetings of the CoP thereafter often appeared keener to blame the criteria rather than their ability to convince their peers through evidence. At the time of their adoption, Parties had agreed that the scientific validity of the new criteria, definitions, notes and guidelines, and their applicability to different groups of organisms should be fully reviewed before CoP12 in 2002. This review was launched in 1999 and quickly developed into a major effort involving the establishment of a Criteria Working Group. Most members of the Criteria Working Group were scientists from Parties selected on a regional basis, but significantly it also included representatives from FAO and ITTO.

The FAO secretariat and some Parties such as Japan, were becoming increasingly concerned about the listing of commercially exploited aquatic species in the CITES Appendices which they saw as a threat to FAO's primacy in the field of managing such species. Surprisingly perhaps for a related UN organization, FAO set up their own review of the CITES listing criteria in 1998 and held two technical consultation meetings on the matter.[34]

When the Criteria Working Group established by the CoP reported to CoP12 in 2002, many Parties felt that they had departed too far from their terms of reference and had not sufficiently reviewed the scientific validity of the existing criteria for a range of species. A further round of work was approved, led by the Animals and Plants Committees, whose efforts resulted in a compromise which received support at CoP13 in 2004. The most significant change to the 1994 criteria, involved the addition of a footnote at the request of FAO, allowing commercially exploited aquatic species to decline to greater levels than other species, before a CITES listing in the Appendices was triggered. This was designed to reflect the greater fecundity of marine species and the ability of their populations to recover following any period of over-exploitation. Numerical guidelines were included, but as examples only as Parties agreed that it was impossible to give numerical values applicable to all taxa because of differences in their breeding biology. It was however left unclear if these guideline figures applied to both of the threat criteria in the resolution which may be used to judge if a species should be included in Appendix II: those species whose trade regulation is necessary to avoid it becoming eligible for inclusion in Appendix I in the near future, and those whose trade regulation is necessary to avoid over-harvesting more generally.

This lack of clarity led to differences of interpretation and significant friction between the FAO and the CITES Secretariat.[35] Further consideration of this point of disagreement rumbled on for a further nine years until CoP16 in 2013. The reality, however, was that CITES Parties clearly wished for a more flexible approach to listing commercially exploited aquatic species, and despite FAO's efforts to limit the Convention's incursion into the regulation of trade in such species, listings continued apace and with an increased degree of support from Parties.

The revised listing criteria agreed in 2004 have been subject to minor modification since, but the principles remain the same. Whilst they have proved a useful guide for Parties, they seem to be quickly swept aside, particularly when more 'media-friendly' species are being considered. Reasons cited by Parties for supporting or opposing a listing proposal often do not focus on the scientific and technical merits of the proposal in relation to the listing criteria, but reflect other concerns. Prime amongst these are socio-economic concerns which are almost completely absent from the listing criteria. Although reference is made to the impact of listing decisions in CoP resolutions about the recognition of the benefits of trade in wildlife[36] and about livelihoods,[37] both refer to the *implementation* of the agreed listing decisions and not to the listing decisions themselves. CoPs have recognized that the range States of a species subject to an amendment proposal should be consulted by the proponent of the proposal. The format for the supporting statements for proposals includes a section where details of the consultation undertaken to secure the comments of the range States concerned should be provided.[38] The importance that should be attached to the views of range States of species proposed for listing is not clear. Many species have been listed against the wishes of some or all of their range States, a situation that still rankles with some Parties.

Efforts to find compromise over contentious listing proposals over the years have led to increasing use of the practice of including annotation footnotes to the listing of species in particular Appendices. These often define the scope of the inclusion of a species in the Appendices, including or excluding different populations of the species and types of specimens, or adding conditionality to the listings. They have had to be accompanied in some cases by explanatory definitions of some of the terms and expressions used in them. Over 130 listings in the Appendices are now subject to an annotation, and their complexity has increased year after year making life complicated for national implementation and enforcement officials and traders alike. The Conference of the Parties asked the Standing Committee to develop proposals to harmonize and streamline the use of annotations, but despite years of effort, this painstaking task has not been completed.

The internal rules for changing the Appendices have consumed huge amounts of intellectual effort and meeting time for the Parties to, and observers of, the Convention. The present criteria are undoubtedly a huge improvement on those applied in the early years of the Convention's implementation, but notwithstanding these efforts to promote decision-making based on objective application of the criteria devised by the Parties themselves, there is an increasing trend to ignore the listing criteria when amending the Appendices. Ideally, Parties should amend the listing criteria in order for them to reflect better their interpretation of the meaning of the text of the Convention. However, Parties who are broadly supportive of wildlife trade and Parties who are broadly against fear that opening up Pandora's box may result in an outcome with which they disagree

4.3   AMENDING THE APPENDICES

Over the years, decisions over the lists of species included in the Appendices of the Convention have created more drama than any other aspect of the Convention's implementation. Arguably, the publicity surrounding these changes has been the

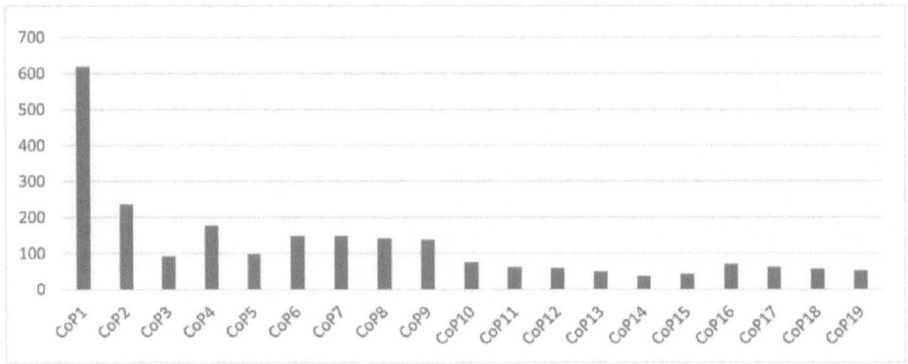

FIGURE 4.2   Number of proposals to amend the Appendices submitted at each meeting of the CoP.

single biggest factor in generating interest in CITES from the general public. The punctual nature of these decisions (at each meeting of the Conference of the Parties), the lead time before the decisions are taken (allowing an 'atmosphere' to develop) and the way that contested matters are decided (often by vote) creates the perfect mixture for an increasingly mediatized world.

It took time for the contents of the CITES Appendices to settle down. CoP1 in 1976 was faced with no less than 619 proposals to amend the Appendices, but from CoP10 in 1997 onwards the number of proposals considered at each CoP has stabilized at just over 50 proposals per CoP (Figure 4.2).

Most proposals submitted to CoP have been to add new species to the Appendices or to increase the level of controls applied to international trade. In part, this reflects a general deterioration in the status of biodiversity and consequent increase in the significance of international trade in the conservation of species. It is also a reflection of the political balance at meetings of the CoP: wealthier countries, which are broadly more likely to be in favour of trade restrictions, have greater resources to prepare written proposals—the thoroughness and technical standard of which has been raised considerably. The trend also reflects the influence of Western conservation and animal welfare NGOs who exert increasing pressure on Parties to initiate or support increased trade controls at CoPs.

Prior to CoP consideration, consultations about the listing proposals are also made with certain other intergovernmental bodies in relation to amendment proposals. The text of the Convention itself requires the Secretariat to consult intergovernmental bodies having a function in relation to all marine species, especially with a view to obtaining scientific data that these bodies may be able to provide and to ensuring co-ordination with any conservation measures enforced by such bodies. When the Convention was negotiated, this was mainly directed at consultations with the International Whaling Commission, but as the species subject to listing proposals have changed, it now relates mainly to FAO and regional fisheries management organisations and regional fishery bodies. Within FAO, an initiative was agreed in 2002 in response to the difficult discussions with CITES over its listing criteria as they applied to commercially exploited aquatic species. FAO's Sub-Committee on Fish Trade of the Committee on Fisheries proposed terms of reference for an ad hoc expert advisory panel for assessment of proposals to CITES. At the meeting which agreed this move, many members [of the Sub-Committee] expressed their strongly held view that CITES listings of commercially exploited marine species should be limited to exceptional cases because, in their view, such listings have potentially serious negative consequences for normal fishing activities, particularly those of developing coastal States including small island States and their economies.[39] The proposal from the Sub-Committee on Fish Trade was endorsed by the FAO Committee on Fisheries, and the first FAO panel meeting was convened to review all proposals to amend the Appendices concerning commercially exploited aquatic species submitted to CoP13 in 2004.[40] Panel meetings have been held before each CITES CoP since then.[41] FAO panel meetings have not been provided for in FAO central funding and depend on external funding, mainly from Japan. FAO panels have been comprised of about ten 'core' fisheries expert members, drawn from academia, research institutes, and government ministries. These members have often participated in previous panel meeting. These core members are supplemented with a number of experts on the individual species subject to amendment proposals at the CoP concerned. The CITES Secretariat has participated as an observer. Panel meetings last a week, and the work is exhaustive. Early FAO panel reports were eagerly awaited by Parties who took careful note of them when making decisions at CoPs. However, to the frustration of FAO, less and less attention has been paid to them over time. At CoP19 in 2022, six proposals were made concerning commercially exploited aquatic species. FAO's panel did not consider that any of them met the CITES listing criteria as drafted, but the CoP approved them all regardless, some by consensus.

By resolution, Parties have also decided to call for the views of the International Tropical Timber Organization (ITTO), FAO, and IUCN on any proposal to amend the Appendices in relation to tree species. These views are sought by the Secretariat but are not derived from an expert panel like those of FAO. In fact, the responses are often rather perfunctory and do not carry much weight with Parties.

Mention should also be made of analyses of the proposals made to amend the Appendices at each CoP by IUCN and TRAFFIC, which started at CoP7 in 1989. The goal was to bring more objectivity and better science into the CITES decision-making and to facilitate the adoption of rational decisions.[42] The process involves expertise from many specialists in the IUCN Species Survival Commission and, for earlier editions, the UNEP World Conservation Monitoring Centre. The IUCN and TRAFFIC analyses are widely considered to be authoritative and were well-received by Parties, particularly in the earlier years when access to information was more limited than it is today.[43] At various times, they have been given 'semi-official' status, either submitted to the CoP with official documents or supported financially either through the Trust Fund, which first happened at CoP11 in 2000,[44] or using external donor funding channelled through the Secretariat. For most of the time however, they remained independent, funded directly by donors, often European Parties.

Finally, the text of the Convention also calls for the Parties to receive the recommendations of the Secretariat on each proposal to amend the Appendices. In order to maintain complete neutrality, the Secretariat's assessments and recommendations are undertaken entirely in-house and are based on the criteria which the Parties have agreed. They do, however, take fully into account the views of statutory consultees, such as FAO for marine species and ITTO, FAO, and IUCN for tree species, and are informed by the IUCN/TRAFFIC Analyses.

4.4   REVIEWING THE SPECIES ALREADY LISTED

As mentioned earlier in this book, CoP1 in 1976 considered that some species of plants and animals may have been incorrectly placed on the Appendices at the time of adoption of the Convention, if the new listing criteria had been in place. For plants, the CoP recommended IUCN review of this matter, and they subsequently identified 14 species for which downlisting or delisting was recommended. They were neither threatened by international trade nor likely to become so.[45] However, it took many years for most of their recommendations to be agreed by the CoP. At CoP3 in 1981, after noting that many species had been added to the Appendices before any objective listing criteria existed, Canada and Switzerland proposed an extensive review of all the listed species by the end of that year, with a similar review after each ten years—the 'Ten Year Review'.[46] The resulting resolution[47] called for a review of the current trade and biological status of all listed species as a method of evaluating the effectiveness of the Convention. Lack of funds and organizational will prevented this exercise from being conducted as planned, and it was eventually wound up at CoP9 in 1994 after just 131 proposals to delete species from the Appendices had been presented to CoP by the Swiss Depositary Government—not all of which were agreed by the CoP. When the Animals and Plants Committees were established at CoP6 in 1987, they were charged with undertaking a 'Periodic Review' of the Appendices to evaluate whether species were appropriately listed.[48] For a while they continued to prepare proposals under the Ten Year Review banner, but from CoP10 in 2000 onwards the resulting proposals to amend the Appendices were submitted to CoP under the title Periodic Review. An elaborate process was agreed to conduct this Periodic review at CoP14 in 2007,[49] but again the results have been modest. Only 47 proposals to amend the Appendices have been made, and again, not all of these were approved by CoP. Overall, the process barely made a dent in establishing compliance with the new criteria for the 40,000 species listed in the Appendices, and it was a considerable workload for the science committees. Most of the species deleted from the Appendices had rarely, if ever, been in international trade, and although they did not merit listing, their inclusion did not present a burden for the Parties. Used as a house-keeping exercise, as presently drafted, the periodic review process does not appear to be a good use of time and resources.

4.5   REVIEW OF SELECTED SPECIES LISTINGS

4.5.1   WHALES

Campaigns against whaling captured the attention of the Western general public around the same time as CITES came into effect. The first Appendices contained only eight whale species, all listed in Appendix I, with an annotation to indicate that they were already protected under the International Whaling Commission's (IWC) schedule of 1972. Parties in favour of restricting or halting whaling generally sought to align CITES provisions with measures taken or recommended under the IWC, where they were getting the ascendency. Thus, at CoP1 in 1976, the USA successfully proposed that Fin whales *Balaenoptera physalus* and Sei whales *Balaenoptera borealis* should be listed under CITES with stocks protected by IWC measures included in Appendix I and the rest of the species in Appendix II.

The Special Working Session of the CoP in 1977 requested the Secretariat to establish closer relations with IWC. The USA sought to tighten the link between CITES and IWC. At a Special Meeting of the IWC held in Tokyo, Japan, in December 1978, they proposed that IWC request CITES CoP2 in 1979 to take all possible measures to support any ban on commercial whaling for certain species and stocks of whale species that IWC had decided[50] were endangered. The decision was carried at the IWC meeting with five votes in favour, four against, and seven abstaining, which would prove important to both CITES and IWC. At CoP2 in 1979, this request from IWC was turned into a CITES resolution which recommended that CITES Parties join the IWC and do not allow commercial trade in specimen of species protected from commercial whaling by IWC, with the lists of those species to be circulated to CITES Parties regularly.[51] At the same CoP, the UK successfully proposed that all cetaceans be listed in Appendix II. Their supporting statement for this proposal did not clearly address the adopted listing criteria for all species but stated that the history of cetacean exploitation had shown that unregulated harvesting always results in population collapse and that all steps should be taken to prevent any other species entering the cycle of discovery, exploitation, over-exploitation, and final collapse.[52] Further IWC-protected species were listed in Appendix I at CoP2 in 1979, CoP3 in 1981 and CoP4 in 1983. Such listings were also seen by proponents as a way of binding CITES Parties to IWC protection decisions even if they were not an IWC member.

IWC introduced a temporary moratorium on commercial whaling, in 1986. Norway and Japan entered an objection to this decision and, with a number of other Parties, also exercised their right to enter a reservation against CITES listings related to 'great whales'. At CoP9 in 1994 a counter-current emerged as Norway proposed the transfer of the Northeast Atlantic and North Atlantic Central stocks of Minke whale *Balaenoptera acutorostrata* from Appendix I to Appendix II.[53] The proposal was soundly defeated in a vote, with 16 Parties in favour and 48 against. Norway presented the same proposal at CoP10 in 1997, where, after a secret ballot, they achieved a simple majority, but not the two-thirds required to carry the proposal. Norway made a third attempt at CoP11 in 2000, when the proposal was again defeated with 52 in favour and 57 against. Norway did not try again, but as permitted under CITES, they exported meat from some of their catch of the species to Japan as both Parties had entered CITES reservations. Japan also attempted to reopen international trade in great whales, with downlisting proposals submitted from CoP10 in 1997 to CoP13 in 2004 for specific stocks of minke whales, grey whale *Eschrichtius robustus* and Bryde's whale *Balaenoptera brydei*. Each time their proposals failed to achieve the two-thirds majority required to be adopted or were withdrawn in the face of opposition. The heated debate over these proposals took a great deal of conference time and energy, and over time support for their proposals showed signs of waning. Japan did not make further proposals after CoP13 in 2004. The co-ordination of conservation measures between CITES and IWC was maintained and indeed strengthened at CoP11 in 2000, when, after a suggestion from the USA, the dispersed policy of the Parties on whaling was consolidated into a single resolution.[54] Whaling ceased to be a feature on the agenda of CITES meetings after CoP13 in 2004, with those Parties engaged with the issue seemingly concluding that they had achieved the best outcome from CITES that they could. With whales largely off the agenda, the Convention could focus on other issues, some of which were equally high-profile and time-consuming.

4.5.2 African Elephant

The CITES logo featuring an elephant was developed for CoP3 in 1981 held in New Delhi, India, but ironically it was not the endangered Asian elephant *Elephas maximus* that was to be the dominant species in the history of CITES, but its relative the African elephant *Loxodonta africana*. Discussions about African elephants in CITES have at times appeared endless, and many publications have appeared on the subject.[55] They provoked and continue to provoke fierce debate at CoPs, exposing some of the more fundamental policy fault lines between Parties in the operation of the Convention. This brief overview broadly summarizes the way that the Parties have addressed the regulation of trade in African elephant specimens, particularly their ivory, but the significance of African elephants, both practical and symbolic, in the development of CITES cannot be understated.

African elephants were not mentioned in the first Appendices of the Convention, but when Ghana joined the Convention on 12 February 1976, they requested that the species be included in Appendix III. At CoP1 in 1976, Canada proposed that African elephants be included in Appendix I, but withdrew their proposal and instead, the species was included in Appendix II as proposed by Switzerland. The Swiss proposal was agreed 'after considerable discussion',[56] but in the absence of any agreed format for listing proposals, this was based on the briefest of written justification:

> Though the African elephant is not threatened with extinction nowadays, the control of the trade, (mostly illegal), however is absolutely essential.[57]

The price of African elephant ivory had risen spectacularly in the early 1970s,[58] perhaps linked to the global recession 1973–1975, and concerns were emerging about the impact of increased demand and trade on elephant conservation. Two reports were to prove significant in raising the profile of the issue in CITES. In 1976, the IUCN initiated a three-year project to investigate the status and distribution of the African elephant, with the aim of drawing up a comprehensive programme for its conservation. The study concluded that the total population of African elephants was 1,343,340 animals.[59] Despite the fact that this total evolved from a good deal of estimation and extrapolation, it became a benchmark figure against which later surveys were compared. The second, parallel study, was of the international ivory trade, undertaken by IUCN and funded by the USA. The US Fish and Wildlife Service had classified the African elephant as a 'threatened' species for the purposes of the US's Endangered Species Act in 1978, and new regulations were imposed which tightened restrictions on the trade in ivory in the USA. The terms of reference of the IUCN ivory study were as follows:

To quantify the world trade in elephant ivory;

To estimate the world-wide investment in elephant ivory and its products;

To assess the role of elephant ivory as a currency equivalent;

To describe the component links in the economic chain of the ivory trade;

To estimate the impact of the ivory trade on the survival of wild elephant populations;

To recommend means for the regulation of the trade to lessen any adverse impacts.[60]

It concluded that although the African elephant as a species is not endangered, it was threatened locally and regionally by the ivory trade, and stronger measures were needed to regulate this trade.[61]

The first meeting of CITES Technical Expert Committee on Harmonization of Permit Forms and Procedures discussed the issuance of CITES permits for ivory in January 1980[62] and as a result a resolution was submitted to CoP3 in 1981. The Parties contented themselves largely with restating the obligations of the Convention in relation to trade in African elephant ivory.[63] They did however introduce a recommendation that each tusk or piece of raw ivory in trade be marked in a standardized way, the details of which arose from a contract that the Secretariat had placed with the Association of European Ivory Traders. Discussions about how to better control trade in African elephant ivory continued in the Technical Committee, but progress was rather slow and not helped by the fact that a number of key African States trading ivory were not yet Party to the Convention. CoP5 in 1985 established an annual export quota system for raw ivory to be held on a central database by the Secretariat. Importing Parties should only import raw ivory which fell within these quotas (including from States not Party to the Convention). Parties were also recommended to mark existing stocks of raw ivory for international trade and report the quantities present to the Secretariat. The Secretariat set up an Ivory Unit, funded in large part by the Ivory Division of the Japan General Merchandise Importers' Association, to manage the process. When Germany, as one of the prime movers behind the initiative, declared the adoption of the resolution the greatest success of the meeting, they were greeted with applause.[64] But the euphoria was short-lived. At CoP6 in 1987, after receiving the Secretariat's report on the quota system, Germany expressed concern that the same resolution was being used as an instrument to allow illegal ivory to re-enter trade.[65] The Secretariat was criticized, particularly in relation to trade in ivory 'exported' from Burundi, which at the time was believed to possess only one remaining elephant—in its national zoo. IUCN presented a rather alarming report to the CoP[66] suggesting a 33% decline in African elephant populations between 1981 and 1987. The tusks of an estimated 89,000 elephants were said to have been traded illegally in 1986, aided by ineffective implementation of CITES in some countries and widespread corruption in some others. Parties responded with a barrage of six new resolutions. Amongst the new measures were the following:

- Action against Burundi and the United Arab Emirates who were believed to be playing a significant part in illegal trade ivory.
- The establishment of a Standing Committee African Elephant Working Group to more closely supervise the Secretariat's administration of the export quota system.

- Exploration of sources of revenue to finance enforcement and the Secretariat's Ivory Unit.
- Stricter controls on worked ivory.
- Registration/licensing of commercial importers and exporters of raw ivory.

A huge public media campaign developed around the theme of 'save the elephants', culminating in the ceremonial burning of 12 tonnes of ivory by Kenya on 19 July 1989. Numerous new elephant NGOs were created, particularly in Western countries. Under pressure from their citizens, many Parties, including the USA and European States, instituted their own national bans on international trade in ivory.

At CoP7 in 1989, the Secretariat provided a lengthy defence of its Ivory Trade Control System,[67] which it claimed had reduced the quantity of ivory being recorded in international trade. However, there were plenty of signs that they had been overwhelmed by the task. The head of their Ivory Unit had resigned, and they had experienced a significant lack of cooperation from Parties. There was no core funding for the Ivory Unit's operation, two-thirds of the costs were provided by ivory trade organizations, a quarter by Parties and less than 10% from NGOs who had been waging a huge media campaign to draw attention to the plight of African elephants. Austria, Gambia, Hungary, Somalia, Kenya, Tanzania, and the USA proposed uplisting the species to Appendix I. A number of Parties in Southern Africa who used the ivory trade as a source of revenue and whose elephant populations were largely unscathed by the wave of illegal killing opposed the proposal as unfairly penalizing their conservation management success. After four days of feverish debate, a compromise proposal from Somalia[68] was agreed whereby the transfer to Appendix I occurred, but a 'Panel of Experts' was to be constituted to provide guidance to Parties on proposals to return some national populations of the African elephant to Appendix II at future CoPs. There was a sense of euphoria amongst those that had campaigned for a ban on commercial international trade in ivory.

A Panel of Experts was duly established and reported on the three proposals made to CoP8 in 1992 to downlist the African elephant populations of Botswana, Malawi, Namibia, South Africa, and Zimbabwe. Although the Parties concerned offered to maintain a moratorium on ivory exports until CoP9, there was little support for their proposals. As the proponents withdrew them, the Minister of Commerce and Industry of Botswana claimed that having satisfied all the required criteria, the goalposts had now been moved. He continued:

> There is a bigger question which we will now have to address—that of evaluating the costs and benefits of remaining within this treaty. We will review our participation in CITES as soon as we have reported to our respective governments. We assure you that our evaluation will be conducted in an objective and analytic manner without allowing emotion to cloud our judgement. Our decision will be in the interest of elephant conservation and the long term benefits to the people of the region.[69]

Sensing that they may be in for disappointment over their elphant proposals, at the same conference Botswana, Malawi, Namibia, and Zimbabwe put forward a proposal to list herring (*Clupea harengus*) in Appendix I[70] specifically noting that range States of the species had not been consulted in advance. The proposal had the desired result with European countries fishing for the species scrabbling to assemble a defensive briefing for the expected discussion. In withdrawing the proposal before it was debated, Zimbabwe remarked that like the African elephant, the solution to the conservation of the species was the development of improved management for sustainable use.[71]

The scene was set for a bitter and prolonged battle. CoP10 in 1997 was to be held in Zimbabwe, and those Southern African Parties advocating the resumption of some international trade in ivory and other Africa elephant specimens sensed their opportunity. Botswana, Namibia, Zimbabwe proposed the transfer of their populations of African elephants from Appendix I to Appendix II and the export of various elephant specimens. Included amongst them was the export, during 1998 and 1999, of their registered stocks of ivory to Japan. After a long and heated debate, amended versions of their proposals were agreed by secret ballot.[72] The result was greeted by singing amongst some of the African observer organizations present, but with anger by Western NGOs who felt that they had been denied sufficient opportunity to speak by the chair of the sessional committee. The implementation of the ivory component of the package was subject to final approval by the Standing Committee.[73] This approval was forthcoming in February 1999, and the ivory was sold by auction in April 1999 with the 49.4 tons sold, representing the tusks of around 2,700 elephants. This raised about 50 million USD for enhanced elephant conservation, monitoring, capacity building, and local community-based programmes. An important component of this package was the establishment of two monitoring systems, under the supervision of the Standing Committee to record illegal hunting of and trade in elephant specimens. The primary objective of their establishment was to assess whether and to what extent observed trends of illegal hunting and trade in ivory were a result of changes in the listing of elephant populations in the CITES appendices and/or the resumption of legal international trade in ivory.[74] These systems became known as the Monitoring the Illegal Killing of Elephants (MIKE) and the Elephant Trade Information System (ETIS). They became fully operational in 2001. Later their objective was expanded to include building capacity in managing elephants and enhancing enforcement. ETIS was and continues to be run by the NGO TRAFFIC, and MIKE was established by the Secretariat. The initial blueprint was coy about the cost of these exercises and the source of funding, speaking only of 'substantial funding' being needed. Examination of reports to CoPs since CoP11 in 2000 shows that the combined cost of both of these systems has run to tens of millions of US dollars, provided mainly by the European Commission/European Union. MIKE organizes patrols at sample sites covering the range of the African and Asian elephants which count the number of elephants carcasses, using their assessment of the proportion killed illegally as an indicator of poaching pressure over time. ETIS morphed into a tool used primarily in compliance-related work targeting Parties with unregulated internal markets for ivory or with significant levels of illegal trade in ivory, in the context of a National Ivory Action Plan Process[75]

After the transfer of elephant populations in Botswana, Namibia, Zimbabwe to Appendix II at CoP10 in 1997, South Africa's population was downlisted from Appendix I to Appendix II with similar conditions at CoP11 in 2000.

CoP12 in 2002 sanctioned a further conditional sale of ivory stocks by Botswana, Namibia, and South Africa using an increasingly complex set of safeguards in the form of conditions attached to the downlisting. In addition to Japan, other importing countries were permitted to purchase the ivory if the Standing Committee was satisfied that they had sufficient national legislation and domestic trade controls to ensure that the imported ivory would not be reexported and would be managed in accordance with all other CITES requirements which included licensing of all importers, manufacturers, wholesalers, and retailers dealing in ivory products. The acceptance of the proposals at CoP12 in 2002 meant that a rival proposal by India and Kenya to return all the Southern African populations of the African elephant to Appendix I could not be discussed.[76]

It took some time for all the conditionality clauses to be met, with both Japan and China eventually designated as trading partners who met the required conditions. It was only in October/November 2008 that just over 105 tonnes of ivory were sold by auction in the four countries, raising just under 15.5 million USD for elephant conservation and for human community development programmes in areas where elephants are found.[77]However, in the meantime, the wind of change was affecting legal ivory exports. At CoP14 in 2007, India and Mali proposed that the four Southern African countries should not be allowed to trade further in raw or worked ivory for 20 years to allow elephant populations time to recover from the effects of ivory trade, and to implement monitoring, enforcement, and control measures.[78] During lengthy negotiations, this was amended, and an agreement made that no further proposals to allow trade in elephant ivory from populations already in Appendix II should be submitted to CoP until November 2017 (nine years from date of the second sale). In the meantime, the Standing Committee was instructed to develop a 'decision-making mechanism for a process of trade in ivory under the auspices of the Conference of the Parties'.[79] Essentially, this was understood to mean the circumstances under which Southern African Parties could resume international trade in ivory and how a decision to allow such trade should be taken. Years of wrangling in the Standing Committee followed with no mechanism developed. Eventually, at CoP17 in 2016, after a vote with 20 in favour and 76 against, it was decided not to continue with the mandate to the Standing Committee to develop such a mechanism.[80] Ironically, because they have not been amended to reflect this decision, the Appendices of the Convention still state that any proposals to allow trade in elephant ivory from the non-endangered populations of Botswana, Namibia, South Africa, and Zimbabwe should be dealt with in accordance with the decision-making mechanism that the Standing Committee was instructed to develop in 2007. Rather like the response to the Panel of Experts report at CoP8 in 1992, Southern African Parties could be forgiven for assuming that they were chasing a mirage.

There were though signs of some common understanding as well. All African elephant range States agreed there was a need for an African elephant action plan to improve elephant management more broadly.[81] The Plan appeared at CoP15 in 2010,[82] and the Secretariat was instructed to establish an African Elephant Fund to support its implementation. This was launched in 2011,[83] with a target of raising USD 100 million by 2014 to enhance law enforcement capacity and secure the long-term survival of African elephant populations. Its administration was later transferred to UNEP which had already established an Elephant and Rhinoceros Conservation Facility with similar goals. However, funds did not follow, and 13 years later only USD 5.5 million had been raised, mostly from European States. The USA already had its own African Elephant Conservation Fund. Despite endorsement of the plan by the Convention on the Conservation of Migratory Species of Wild Animals in

2017 and a revised version for 2023–2027 being agreed,[84] the initiative has simply not attracted enough support to make a difference to elephant conservation at a continental level.

During the ten-year pause in trade, a vigorous public debate developed about whether the two sales of stocks had exacerbated illegal trade in ivory or rewarded the benefitting Parties for their successful elephant conservation and management. Marshalled by Western NGOs,[85] an independent African Elephant Coalition (AEC) was established in 2008, comprised mostly of the States that had lost many of their elephants to illegal killing for the ivory trade. They adopted the 'Bamako Declaration' opposing all international trade in ivory:

> The coalition will strive to have a viable and healthy elephant population free of threats from the international ivory trade. Parties to the coalition will also develop an elephant action plan that will encompass national and regional elephant strategies that promote non consumptive use of elephants through development of ecotourism for the benefit of local communities.

At subsequent CoPs, attempts by Tanzania and Zambia to emulate other Southern African Development Community States and have their populations downlisted to Appendix II and to sell their ivory stockpiles were unsuccessful. Members of the AEC tried to tighten existing controls but withdrew them as a stalemate developed.

As the close of the nine-year moratorium arrived, at CoP17 in 2016, Namibia and Zimbabwe proposed that their population of African elephants be maintained in Appendix II, but unconditionally, to allow international trade in their ivory to be largely unfettered. Both proposals were defeated in a vote, but so too was a proposal from a number of AEC Parties to return the whole species to Appendix I. Similar proposals from both 'sides' were made to CoP19 in 2022 with the same result. And so the deadlock continued. Controversy in recent years has moved from ivory to the subject of international trade in live young elephants. The 1997 downlisting of Botswana, Namibia, Zimbabwe elephant populations had permitted the export of live elephants to 'appropriate and acceptable' destinations. This phrase had been borrowed from earlier permission for South Africa to sell live white rhinoceroses (*Ceratotherium simum*), where it meant destinations where their horns would not subsequently be removed. Its application to African elephants was never really questioned until Southern African States found a lucrative market for live elephants in China for use in zoos and circuses. This was a useful source of revenue to replace some lost from the ivory trade ban, but quickly became a campaigning theme for NGOs.[86] In the context of African elephants, the phrase 'appropriate and acceptable destinations' was subsequently defined by CoP to mean for use in in situ conservation programmes or other places where demonstrable in situ conservation benefits for African elephants would accrue. Namibia found a way around this by treating the live elephants involved as if they were in Appendix I where the definition did not apply. An arcane debate continues to occupy plenty of the Parties' time, despite the number of specimens involved, a few dozen per year, having no direct impact on the conservation of African elephant populations. As a CITES Secretariat media release carefully said, this is 'a very sensitive issue that generates expressions of public concern'.[87]

The result of the MIKE monitoring showed that elephant poaching pressure generally increased between 2003 and 2011 but decreased thereafter. The monitoring concluded that whilst there was a strong positive association between the global annual trend in the price of elephant ivory and poaching pressure, there was no evidence to suggest that illegal killing of elephants increased or decreased as a result of the sales of ivory that had been permitted by CITES or the nine-year moratorium on those sales.[88] Rather, at the supply end, evidence suggests that illegal killing of elephants tends to be lower in countries with better governance quality, at monitoring sites with higher law enforcement capacity, and where households adjacent to areas where elephants are killed are wealthier and healthier.[89] Changes in legislation and consumer preferences in China,[90] as the most significant market for ivory, were probably a major factor in the reduction of demand.

The most authoritative source of information on the number of African elephants found in the wild is the African Elephant Database, maintained by IUCN from which periodic African Elephant Status Reports have been produced and submitted to CITES CoPs. Considering the amount of time and effort spent discussing elephants and ivory trade controls in CITES, estimates of African elephant population sizes and trends are remarkably uncertain. Despite survey techniques being improved and resources increased, a significant proportion of the population estimates are still based on speculative expert assessments rather than on detailed survey results. The continent-wide population estimates from the African Elephant Status Reports produced over the years have varied as a result of new and improved knowledge (Figure 4.3). All the population estimates for 1995–2015 are well below the 'baseline' estimate of

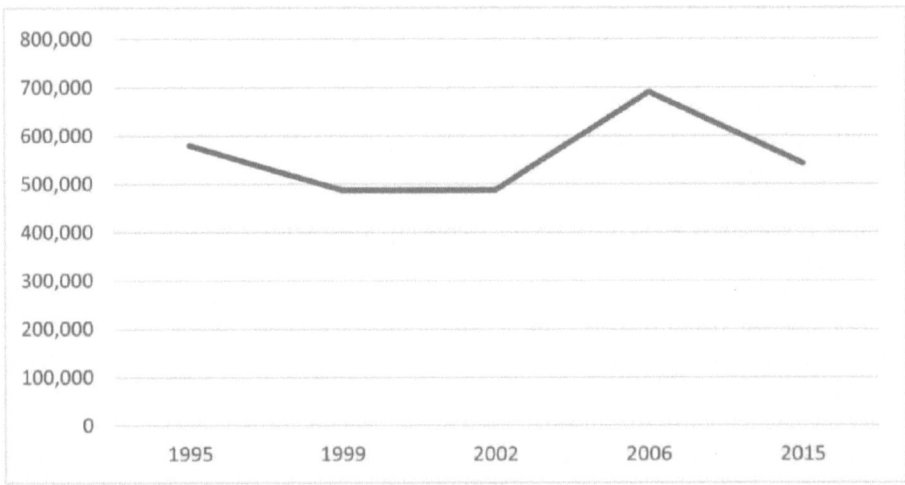

FIGURE 4.3 Continental population estimates for African elephants in published reports from the African Elephant Database[92] (all categories of estimate combined). Median figure used where a higher and a lower estimate are given.

1,343,340 animals which first triggered action by CITES in 1979. However, in these early days, in most parts of Africa, information on elephant numbers was based on guesswork and has not formed a suitable basis from which to determine population trends over time, or to discern the effects of policy or management.[91]

What is clear is that there is great variation in the elephant population trends between different parts of the African continent. Broadly, Southern African populations have remained stable or increased whereas those in other areas have declined, sometimes significantly. Southern Africa now holds nearly 75% of the total continental African population. Many have tried to determine if the actions of CITES have had a positive impact on Africa elephants, but evidence is contradictory, and establishing cause and effect is confounded by the number of variables involved. Except for the MIKE and ETIS analyses, the almost complete absence of any socio-economic background for CITES decision-making over the ivory trade is a particularly striking omission. Whilst some academic studies addressed this point, for the most part, CITES policies did not. The focus on African elephants in CITES undoubtedly precipitated domestic trade restrictions or bans in most market countries: particularly China, Japan, the European Union, and the United States of America. CITES has not proved sufficiently flexible to allow Southern African States whose elephant management model is clearly a success, to continue with their policy of utilizing elephant ivory. Whilst these countries have repeatedly threatened to leave CITES unless a more rational approach is adopted, they have never done so. The reality is that with most markets for ivory closed and public opinion, particularly in Western countries resolutely opposed to any consumptive use of elephants, sustainable or not, these States would have few markets to sell their ivory into, even if they left the Convention.

In 2021, a scientific consensus emerged that African elephants were in fact two species; African forest elephant *Loxodonta cyclotis* and African savanna elephant *Loxodonta africana*.[93] CITES Parties are now considering whether to adopt this change and, if so, what impact it would have on the various policy initiatives taken for African elephants under the Convention. In past years, when there was some ivory trade allowed and the prospect of more being permitted, this change would have presented an extremely difficult problem. As it is, with most international ivory trade shut for the foreseeable future, the question is likely to be more academic.

4.5.3 CROCODILIANS

Alligators, caimans, and crocodiles, known as crocodilians (*Crocodylia* spp.) are regularly portrayed as one of the biggest CITES success stories.[94] A significant trade in American alligator *Alligator mississippiensis* developed in the mid-1880s during the American Civil War to provide shoe and boot leather. Subsequently, and particularly from the 1920s, alligator and crocodile leather became fashionable, especially in Europe, and an international trade developed with wild skins supplied from the world's equatorial regions. At its height in the 1950s and 1960s, before any significant crocodile farming, the global harvest was probably providing about 6–8 million skins per year.[95]

Overhunting of many species became obvious, when the global supply of the most desirable skins declined precipitously. All 22 species of crocodilians then recognised were included in the Appendices when the Convention began, 18 species (or subspecies) in Appendix I and the remainder in Appendix II. Crocodile farming or ranching was embryonic at that time, largely restricted to Siamese crocodiles *Crocodylus siamensis* in Thailand. However, the scarcity of crocodiles and trade restrictions arising from CITES controls resulted in farming being developed for other species, namely, Nile crocodile, *Crocodylus niloticus* (in Zimbabwe); saltwater crocodile, *Crocodylus porosus* (in Australia, Indonesia, and Papua New Guinea); American alligator (in the USA); spectacled caiman *Caiman crocodilus* (in Venezuela); and New Guinea crocodile, *Crocodylus novaeguineae* (in Indonesia and Papua New Guinea).[96]

Crocodilian farming relies on closed-cycle, or largely closed-cycle captive breeding, whereas ranching refers to the rearing in a controlled environment of animals taken as eggs or juveniles from the wild, where they would otherwise have had a very low probability of surviving to adulthood. Crocodiles are often not popular with people living within their range because of the threat that they pose to both human life and livestock. The harvesting of eggs for ranching usually involves local people who can therefore derive some economic benefit from these animals which are otherwise often destroyed when found. At CoP2 in 1979 it was recognised that ranching could be applied for some populations of Appendix I species that could withstand a certain level of exploitation.[97] A resolution on downlisting for ranching purposes was subsequently agreed at CoP3 in 1981. This allowed, subject to certain conditions, consideration of the downlisting to Appendix II of populations of species included in Appendix I which were no longer endangered and whose conservation could benefit from ranching.[98] From CoP4 in 1983 to CoP10 in 1997, 28 proposals were made to ease the strict protection of certain populations of crocodilians under this ranching provision. Proposals for Nile crocodile *C. niloticus*, saltwater crocodile *C. porosus,* black caiman *Melanosuchus niger,* broad-snouted caiman *Caiman latirostris*, and American crocodile *Crocodylus acutus* were approved by CoP. A total of 15 Parties have successfully requested such downlistings for crocodilians, although new requests have lessened in recent years, either because most interested Parties have already taken advantage of the possibility to make such requests and/or the market for certain crocodilian products has been saturated. With the status of a number of crocodilian species in the wild either improving or not being as serious as first believed, a number of Parties took advantage of another process agreed by Parties at CoP5 in 1985. This allowed national populations of species such as crocodilians, which had been included in Appendix I before listing criteria were agreed could be transferred to Appendix II subject to the establishment of an export quota.[99] Between CoP6 in 1987 and CoP9 in 1992, about 30 proposals for crocodilian species were approved by the Parties under this process. In order to try and ensure that these exceptions did not lead to illegal trade, a system of tagging of crocodilian skins and substantial parts of skins was established at CoP8 in 1992[100] and remains in force today. During all these developments, exporting Parties were greatly assisted by the Crocodile Specialist Group of IUCN's Species Survival Commission who provided extensive support both in-country and in CITES conference halls. For Nile crocodiles *C. niloticus* in Africa a key Secretariat project on the species also high-lighted crocodile conservation.[101]

Most crocodilian species, including many of those in international trade, are now not considered threatened with extinction by IUCN in its Red List of Threatened Species. With the advent of farming and ranching the number of crocodilian skins in international trade increased and in the last ten years, average trade has been around 1.5 million skins per year, although it is now showing a gradual decline.[102] The measured reopening of international trade in specimens of crocodilian species is generally considered to have benefited their conservation.[103] A review by IUCN submitted to the CITES Animals Committee in 2006 concluded that ranching of crocodilians, in particular, provides specimens for international trade on the one hand, and creates commercial incentives to conserve adult crocodilians in the wild on the other. It further noted that crocodilian ranching has worked successfully in a variety of different countries with varying socio-economic levels, technical capabilities, and crocodilian species and that nowhere has ranching been associated with, or alleged to be the cause of, detrimental effects on wild populations.[104] However, use for international trade from farms has been implicated in the decline of some species. For example, Siamese crocodile *C. siamensis* is a critically endangered species which has always been and remains listed in Appendix I. The species is however very widely kept on farms in southeast Asia and harvesting from the wild to provide breeding stock for these farms is considered to have been one reason for the species' decline in some countries.[105] The loosening of controls on international trade in crocodilian species has been facilitated by the fact that they lack the more charismatic attraction of species such as sea turtles and elephants. Consequently, whilst the use of crocodilian products and the way that they are produced has its critics,[106] discussions over the matter at CITES meetings have been largely spared the media hype and tension that have characterised debate over more media-friendly species.

4.5.4 Vicuña

Vicuñas *Vicugna vicugna* are wild camelids from the high Andes in South America. Their wool is exceptionally fine and is consequently highly valued, both locally and internationally. They were of great significance during the Inca Empire, and killing of these animals was prohibited. Being hard to domesticate, wild animals were instead captured, sheared, and released.[107] After European colonization replaced the Inca Empire, the very high value of the wool of the species in international markets and local culling for food and to prevent competition with livestock led to a dramatic decline in vicuña populations. A population of perhaps two million animals declined by the mid-1960s to less than 10,000 and possibly as few as 5,000 specimens. Although the subsequent recovery of the species is often portrayed as a success story due to CITES, in fact efforts to restore the species began in Peru, its last stronghold, about 1966, before CITES was agreed.[108] The importance of international trade in products of vicuña and the need for cooperation led to the signing of the Convention between the Government of the Republic of Bolivia and the Government of the Republic of Peru for the Conservation of the Vicuña, which came into force in 1970. In 1979 it became the Convention for the Conservation and Management of the Vicuña and was signed by all the range States of the species.

When CITES came into force, the species was included in Appendix I. By CoP2 in 1979, populations had already risen sufficiently for Peru and Chile to propose a downlisting for their own populations,[109] with their numbers claimed to be 60,000 and 4,000 respectively. However, some thought this reopening of trade was too soon, including Peruvian diplomat, conservationist and self-styled 'Mr. Vicuña',[110] Felipe Benavides Barreda, who represented the International Fund for Animals Welfare at CITES meetings. He led a high-profile media campaign against the move, and the proposals were defeated in a vote. Peru and Chile worked in the forum of the Technical Committee to find a formula for downlisting proposals that might win support. By the time that they came back with revised proposals eight years later, their vicuña populations had risen to 90,000 and 12,000 respectively. Their revised proposals were more nuanced, being limited to particular populations of the species in the two countries and to international trade in cloth made from wool sheared from live vicuñas or items made from such wool—using capture techniques derived from those used by the Incas. In a rather novel development, the reverse side of the traded cloth was to carry a logo adopted by Convention for the Conservation and Management of the Vicuña and the selvages of the cloth was to state either VICUÑANDES-CHILE or VICUÑANDES-PERU, depending on the country of origin. Their proposals were approved by consensus at CoP6 in 1987. With improved control of international trade, the species' status continued to improve: IUCN's Red List rated it 'Lower Risk/conservation dependent' in 1996, moving to 'Least concern' in 2008.[111] A series of 16 downlisting proposals were made to CITES CoPs, both to move populations of the species in different range States to Appendix II, and to adjust the annotation containing the conditionality under which trade was allowed. By 2023, the total population of vicuñas numbered over half a million animals,[112] and all populations in Bolivia, Ecuador, and Peru were listed in Appendix II together with most populations in Argentina and Chile. At CoP17 in 2016, the extraterritoriality of the conditionality measures was extended such that all cloth and garments made from live-sheared vicuña fibre in international trade (included that from non-range States) should be marked with a logo (Figure 4.4).

The conditionality measures were designed to try and ensure that range States, and in particular local people within the range of the species, captured a bigger part of the economic value of the products. Notably, all of these downlisting proposals have been agreed by consensus. Concurrent with amendments to the CITES Appendices for vicuña, a series of resolutions have also been adopted, designed both to exhort Parties—both range Sates and non-range States—to improve their controls

FIGURE 4.4   Logo required for all cloth and garments made from live-sheared vicuña fibre in international trade.

on vicuña products and to strengthen cooperation and coordination between CITES and the Convention for the Conservation and Management of the Vicuña.[113]

Some illegal trade continues, as does grazing competition between vicuña and domestic animals in certain areas. Whilst vicuña populations have increased, local people have not necessarily benefitted as much as they might from the trade and remain poor. It has been claimed that this is due to the oligopsony of the international market for vicuña fibre.[114] A single company, owned by a multinational luxury brand, controls a large part of the global trade in vicuña products and some of the production in certain range States.[115] Rewards are certainly great and increasing. Between 2007 and 2016 the volume of global trade in vicuña products (exports from all producer countries) increased 78% by volume with a value in 2016 of approximately USD 3.2 million per annum.[116] In 2023 a single jacket made from vicuña retailed for over USD 20,000.[117] Regulation of international trade in vicuña products has led to a spectacular increase in populations in the wild and considerable scope remains to extend the area over which vicuña are managed and to improve the equity of international trade.

4.5.5 TIMBER AND TREE SPECIES

Plants were neglected when the CITES treaty came into force and timber species even more so. When the Convention began, nine timber species were listed in Appendix I and four in Appendix II. Seven of these were subsequently removed from the Appendices as they were found not to be in international trade.

Interest in using the Convention to regulate trade in timber can be traced to the establishment of the Plants Committee and a key publication by IUCN in 1988.[118] The first meeting of the Plants Committee in 1988 established an Action Group on Timber, but it was not until CoP8 in 1992 that a substantial number of proposals to include timber species in the Appendices were made, largely promoted by Denmark, the Netherlands and the USA. As the Secretariat noted in its comments on the proposals, regulating trade in timber would present some challenges for Parties, and some felt that CITES was not an appropriate instrument to deal with timber trade.[119] Applying the listing criteria at the time to tree species was problematic as they were unlikely to be threatened with biological extinction by commercial logging. From a practical perspective, the practice in the timber trade of selling timber at sea and/or dividing shipments into multiple lots on arrival in a customs-free zone presented documentary complications in the issuance of CITES permits and certificates, as did the identification of wood to species level. Furthermore, questions were raised about coordinating CITES' actions with that of FAO, which had produced its own list of tree and scrub species threatened with extinction or subject to severe genetic depletion in 1986[120] and with the International Tropical Timber Agreement, which had come into force in 1985. In the face of these complications, a number of the CITES listing proposals made to meetings of the CoP were scaled back or withdrawn. The listing of Brazilian rosewood *Dalbergia nigra* in Appendix I and Afrormosia *Pericopsis elata,* small-leaved mahogany, *Swietenia mahogani,* and Lignum vitae *Guaiacum officinale* in Appendix II were agreed at CoP8 in 1992, and CITES engagement with the international timber trade had begun in earnest.

Initiated by Germany, India, Kenya, and the Netherlands, further listings for timber species were agreed at CoP9 in 1995, but not without difficulty and a number of proposals for more commercially valuable species were withdrawn or rejected. The CoP instructed the Standing Committee to establish a temporary timber working group to address the technical and practical problems associated with the implementation of tree listings and to define the relationship with existing international organizations addressing the sustainable use of timber resources. This group met twice in 1995 and 1996 and produced a lengthy report for CoP10 in 2000[121] which resulted in a suite of detailed technical changes to CITES procedures for regulating international trade in timber. This included a Resolution specifically relating to the implementation of the Convention for timber species,[122] which, amongst other things, called for consultations with relevant international timber organizations prior to making proposals to amend the Appendices for such species. The Standing Committee Timber Working Group formally continued to exist until CoP12 in 2002 but never met again. Meanwhile, the focus in CITES switched to big-leaf mahogany *Swietenia macrophylla*, the most valuable and widely traded Neotropical timber tree. Proposals to list this species in Appendix II were rejected at CoP8 in 1992, CoP9 in 1994, and CoP10 in 1997, but after the joint proposal from Bolivia and the USA failed at the latter, a compromise was agreed in which Parties were encouraged to list the species in Appendix III. Brazil offered to establish a working group on mahogany within the framework of the Amazon Cooperation Treaty to discuss the conservation status of mahogany and make recommendations to achieve the sustainability of international trade in the species.[123] Subsequently, Costa Rica (1995), Bolivia (1998), Brazil (1998), Mexico (1998), Peru (2001), and Colombia (2001) listed their populations of the species in Appendix III.[124] These listings were restricted to wild populations of the Americas, excluding plantation timber from non-range States, and applied only to sawlogs, sawn wood and veneers, so excluding plywood and finished products. They did however require all range States to issue CITES documentation for exports of specimens from the populations that they did cover. The Appendix III listings were not without some shortcomings[125] but served to keep the matter high on the CITES agenda of the Parties concerned, such that, by a narrow majority at CoP12 in 2002, the Parties decided to include the species in Appendix II.[126] The Bigleaf Mahogany Working Group, which met once under the auspices of the Amazon Cooperation Treaty and four times under CITES, served as a forum to exchange experiences of the CITES listings and to maintain momentum. The Working Group was later expanded to cover other neotropical trees and placed under the supervision of the Plants Committee. The Plants Committee also pursued improved non-detriment findings for Big-leaf mahogany *S. macrophylla* under the Review of Significant Trade.[127]

Proposals for CITES to regulate the timber trade for other species were continuing to meet some resistance. At CoP14 in 2007, proposals by Germany to list various rosewood and cedar species had to be withdrawn in the face of opposition from range States.[128] Instead, a plan of action for the species was agreed.[129] When some of the same species were proposed for listing in Appendix II at CoP16 in 2013, the mood had changed, and their inclusion was agreed by consensus.[130] Further, at the same meeting a raft of other timber and other tree species were listed in Appendix

II including around 300 species of ebony and rosewood from Asia and Madagascar. The change in attitude was sparked by an enhancement of cooperation between CITES and ITTO. CoP14 in 2007 had given this cooperation a boost by adopting a resolution on the subject,[131] but under earlier encouragement from Parties,[132] the Secretariat had already embarked on a series of capacity-building initiatives in conjunction with ITTO. In 2006 significant funding from the European Union enabled this cooperation to move ahead on a broader scale and during the following ten years CITES capacity-building activities for tree species were undertaken throughout the world.[133] The Secretariats of CITES and ITTO signed a Memorandum of Understanding in 2018,[134] and although cooperation continued and with similar objectives, from 2017–2022 the programme was recast as the CITES Tree Species Programme.[135] It was struggling to attract donor support until, after a pause in activity, it was relaunched at the end of 2023 when Germany offered a 10 million Euro contribution over five years to support sustainable trade in CITES-listed tree species and related forest governance.[136]

During CoP17, CoP18, and CoP19 in 2016, 2019, and 2022 respectively, the tendency with regard to the inclusion of timber species in the Appendices which first manifested itself at CoP16 in 2013 continued and accelerated. Unlike at previous CoPs, many proposals were now submitted or co-submitted by range States. After debate and sometimes adjustment of conditionalities detailed in annotations, most were agreed by consensus or, if pushed to a vote, by overwhelming majorities. By 2024, the number of tree species included in the CITES Appendices was approaching 1000, many of which are used in the timber trade, with a small number traded for medicinal, aromatic, or other purposes. In 2018, the Secretariat joined the Collaborative Partnership on Forests, an inter-organizational partnership to help enhance the contribution of forests and trees to internationally agreed development goals. However, its proposal to develop a strategic resolution on CITES and forests for the Convention[137] did not immediately find favour amongst Parties who, as so often, preferred a more pragmatic approach focussed more closely on the objectives of the Convention. The history of the application of the Convention to timber and tree species shows what a long process a change in focus can be and also how positive incentives, such as the availability of capacity-building, can influence such change.[138] The challenge now is to ensure that the comprehensive infrastructure developed[139] can help secure timber resources in the longer term.

4.5.6   Fish and Other Marine Species

The application of the Convention to marine species has its own specificities in a similar way to those relating to trade in timber and tree species. In addition to trade between States, the Convention procedures must also be applied to introduction of marine species from the sea—transportation into a State from waters that lies beyond national jurisdiction. Parties wrestled for many years with the complications arising from the application of these provisions. The prospect of species from the high seas being listed in the Convention's appendices prompted Australia to propose some common interpretations of CITES' provisions at CoP11 in 2000.[140] Particular problems were the obligations on flag States of vessels on the high seas and the practice of transhipping of fish at sea from one vessel to another. Their proposal was not successful but provoked a long and ultimately useful debate amongst Parties with a significant input from FAO.[141] The matter was only resolved to the satisfaction of all after a proposal for a resolution from the USA was accepted by Parties at CoP14 in 2007.[142]

The text of the Convention already touches on the issue of the management of marine resources by specifically relieving Parties of their CITES obligations in relation to marine species subject to protection measures under another pre-existing treaties. It also requires consultation with intergovernmental bodies dealing with marine species prior to consideration of their listing in the Appendices.

Other than cetaceans, which are considered earlier in this chapter, the initial Appendices contained only a handful of marine species: four species of sturgeon *Acipenser* spp. and the coelacanth *Latimeria chalumnae*, a primitive oddity that is not fished commercially. Early additions to the Appendices were marine invertebrates—black coral (Antipatharia spp.) at CoP3 in 1981, giant clams (Tridacnidae spp.), and some stony corals (Scleractinia spp.) at CoP5 in 1985, with the remainder of the stony corals at CoP7 in 1989, and queen conch *Strombus gigas* at CoP8 in 1992. None of these were of major global economic significance or greatly impinged on existing marine species governance. At CoP9 in 1994, when the USA presented a discussion document on trade in shark parts and products,[143] the tone changed. The USA was quick to point out that it did not believe that CITES should assume the responsibility for management of sharks 'at this time', but in the absence of any other bodies effectively doing so, a taboo seemed to have been broken by the document. After first seeking to have the matter removed from the agenda of the meeting,[144] Japan led the call for FAO and established international fisheries agreements to deal with shark management, rather than CITES. Although they received some support, it was not enough to prevent a resolution being adopted directing the Animals Committee to prepare a discussion paper on the biological and trade status of sharks. The resolution also requested FAO and other international fisheries management organizations to assemble biological and trade data on shark species for a future CoP and to 'fully' inform the CITES secretariat about their work. The combative tone set the scene for a difficult relationship in later years. The Animals Committee's report to CoP10 in 1997,[145] was comprehensive and the resulting decisions of the CoP[146] were quite far-reaching, considering that no shark species were yet listed in the CITES Appendices. In an effort to counter this initiative, FAO became more active and organized a workshop on the conservation and management of sharks, hosted by Japan.[147] This resulted in the adoption of an FAO International Plan of Action for Conservation and Management of Sharks.[148] The USA already began to test the water, so to speak, at CoP10 in 1997 by proposing the inclusion of sawfishes (rays), Pristiformes spp., in Appendix I. The species were not in significant international trade, but the proposal was defeated with 24 Parties in favour and 50 against.[149] When they made the same proposal ten years later at CoP14 in 2007, 67 were in favour and 31 against,[150] demonstrating the change in outlook of many Parties regarding marine species in CITES.

The USA opened up a more general debate at CoP10 in 1997, by proposing the establishment of a temporary Standing Committee working group on marine fish species,[151] modelled on the CITES Timber Working Group. Whilst the US was at pains to stress the primacy of FAO, the draft terms of reference did include the preparation of an analysis of the technical and practical implementation concerns associated with the inclusion in the CITES Appendices of marine fish species subject to large-scale commercial harvesting and international trade. After lengthy debate, positions were rather polarized, and the proposal to establish the working group was defeated in a secret ballot with 49 Parties in favour and 50 opposed.[152]

Although Atlantic sturgeon *Acipenser sturio* had been listed in Appendix II since the start of CITES, CoP10 in 1997 saw the listing in Appendix II of the other 26 sturgeon species at the request of Germany and the USA. Sturgeons are found in northern Asia, North America, and Europe in both marine and freshwater environments, but it is the Caspian Sea and related waters that produced most caviar—unfertilized eggs—and from the most celebrated species. Sturgeon species generally take many years to reach sexual maturity and do not reproduce every year so are considered vulnerable to overfishing. Beluga sturgeon *Huso huso* are the largest freshwater fishes in the world and have been known to reach 1.5 tonnes in weight and seven metres in length. By the time that CITES arrived on the scene, such large specimens were long gone, and the species was rated Endangered in the IUCN Red List. Caviar has been eaten for hundreds of years and has become associated with luxury. International trade in caviar was severely affected by the political turmoil associated with the dissolution of the Union of Soviet Socialist Republics in 1991. Smuggling of caviar increased, and its price and quality declined, and so quantities consumed outside range States increased from around 300 tonnes per year to about 450–500 tonnes.[153] The listing of all remaining sturgeon species came into effect in April 1998, and the Convention expended considerable effort and resources to try and make sturgeon fishing more sustainable and control illegal trade. After a fractious meeting of the Standing Committee in 2001,[154] Caspian littoral States concerned agreed to develop common management plans and catch and export quotas for sturgeon fished from shared stocks. They formed a Commission on Aquatic Bioresources of the Caspian Sea for this purpose. A CITES marking system was agreed for caviar in international trade.[155] The Animals Committee included the species in the Review of Significant Trade and made recommendations for improving the sustainability of the fishing. Numerous meetings were held, later ones belatedly joined by FAO,[156] who went on to support a capacity-building project for the Caspian littoral States,[157] which included stock assessment and sustainable harvesting techniques together with sturgeon hatchery guidelines. Despite incremental improvements in the system of international trade regulation, illegal trade of caviar from the Caspian basin continued, either bypassing the CITES systems or using false labelling.[158] In the face of declining wild stocks, quotas for the export of caviar from the wild were reduced to zero during the period 2008–2010. CITES continued to support efforts to suppress illegal trade,[159] but investors had moved to the establishment of sturgeon aquaculture for the production of caviar. Most of the aquaculture development took place well away from the Caspian Sea, in China, Europe, the USA, and even the United Arab Emirates and Uruguay. Although arriving late in the day, CITES failed to prevent the commercial extinction of wild sturgeon fisheries. Sturgeons largely disappeared from the agendas of CITES meetings and efforts to improve the situation via CITES measures ceased. Incentives for conserving stocks in the wild disappeared, and local fishers lost their livelihoods, whilst businesses based mostly in consumer States reaped the rewards from sturgeon aquaculture.

In the field of elasmobranchs (sharks, rays, skates, and chimaeras), resistance to CITES listings weakened, and CoP11 in 2000 saw the start of a process that would result in 134 of the roughly 1,250 species of elasmobranchs listed in the CITES Appendices by 2024. Rather like timber species, some of the early listings were made unilaterally in Appendix III, starting with the inclusion of Basking shark *Cetorhinus maximus* by the UK in 2000. The early proposals to include shark species in Appendix I or II were often rejected at first. This being the case for Whale shark *Rhincodon typus* (rejected at CoP11 in 2000, but agreed at CoP12 in 2002), Great white shark *Carcharodon carcharias* (rejected at CoP11 in 2002, but agreed at CoP13 in 2004), Porbeagle shark *Lamna nasus* (rejected at CoP14 in 2007 and CoP15 in 2010, but agreed at CoP16 in 2013), and Oceanic whitetip shark *Carcharhinus longimanus* (rejected at CoP15 in 2010, but agreed at CoP16 in 2013). A rolling programme of reviews and reporting in the CITES intersessional committees kept the issue on the radar.[160] Proposals to add shark species to the Appendices were tabled at every meeting of the CoP, most of which were adopted with increasing ease. The influence to the FAO expert panel declined such that at CoP19 in 2022 nearly 100 species of sharks and rays were included in the Appendices by consensus despite the panel concluding that they did not meet the scientific criteria for listing.

Despite previous frictions, FAO and the CITES Secretariat signed a Memorandum of Understanding focussing on commercially exploited aquatic species listed on the CITES Appendices in 2006[161] which has brought some tangible benefits, particularly in terms of mutual understanding of legal frameworks and the provision of legal training workshops.[162] The Secretariat also partnered with UN Conference on Trade and Development in its Blue BioTrade initiative over the use of Queen conch *Strombus gigas* in the Caribbean.[163]

Not all listing attempts for marine species have been successful: Patagonian and Antarctic toothfish *Dissostichus eleginoides* and *D. mawsonii* at CoP12 in 2002, Banggai cardinalfish *Pterapogon kauderni* at CoP14 in 2007 and CoP17 in 2016, Spiny dogfish *Squalus acanthias* at CoP14 in 2007 and CoP15 in 2010, and Caribbean and Smoothtail spiny lobsters *Panulirus argus* and *P. laevicauda* at CoP14 in 2007—all failed to achieve the required two-thirds majority. The highest profile species which the Parties decided not to list in the Appendices was Atlantic bluefin tuna *Thunnus thynnus*, which is highly prized—and priced—in Japanese cuisine. In 2019 a single fish sold for USD 3.1 million (over 11,000 USD per kilo).[164] Largely due to fishing pressure, the species underwent a huge decrease in abundance. At CoP8 in 1992 a proposal was submitted by Sweden to include the western Atlantic population in Appendix I and the eastern Atlantic population in Appendix II,[165] but it was withdrawn after tuna fishing nations Canada, Japan, Morocco, and the USA submitted a draft resolution instead urging the International Commission for the Conservation of Atlantic Tunas (ICCAT) to maintain its efforts to manage the fishery for the species.[166] This confidence in ICCAT however seemed misplaced after an independent report in 2009 noted that the management of the fishery for the species by ICCAT was widely regarded as an international disgrace.[167] At CoP15 in

2010 Monaco proposed to include the whole species in Appendix I.[168] Monaco's proposal sparked huge interest in the media and an intensive lobbying effort both from supporters of the proposal, backed by many NGO observers and opponents from the nations fishing the species, led by Japan. A majority of the FAO Expert Advisory Panel members concluded that the species met the criteria for inclusion in Appendix I, while all did so in relation to Appendix II.[169] During the debate at CoP15 in 2010, Spain proposed an amendment delaying any application of an Appendix I listing pending a review by the Standing Committee to ensure that it was still warranted. Opponents of the proposal, however, took advantage of the relative inexperience of Monaco in CITES fora to use the Rules of Procedure to curtail the discussion. They forced a vote on the proposal before any attempt could be made to further soften its impact by additional conditionality or by amending it to a proposal for inclusion in Appendix II. In a secret ballot, Spain's revised proposal was supported by 43 Parties with 72 against, and Monaco's initial proposal was then firmly defeated with 20 in favour, 68 against with 30 abstentions. NGOs were hugely disappointed, but the impact of the proposal and the focus of attention on the status of blue-finned tuna *T. thynnus* had a dramatic effect. ICCATT increased its management effort, populations grew and catch quotas could be increased.[170] This demonstrated the influence CITES can have on the management of marine resources without actually getting directly involved itself.

Although regulation of international trade in aquatic species, both marine and freshwater, falls within the competence of CITES in terms of avoiding international trade that could threaten species with extinction, any action in these areas brings the Convention into the domain of other intergovernmental organizations set up specifically to address the managed use of such species. Many of these aquatic species are managed by a patchwork of about 50 regional fisheries management organizations and regional fishery advisory bodies with varying degrees of responsibility and legal competence in relation to the sustainable utilization of living aquatic resources.[171] In 1999, FAO established a Regional Fishery Body Secretariats' Network to facilitate information exchange among their secretariats. At the global level sits the FAO Committee on Fisheries, established in 1965 to review fisheries issues and negotiate binding agreements and voluntary instruments concerning fisheries. Ensuring coordination between CITES and these bodies means negotiating not only institutional 'turf wars' between organizations but also differences in culture. Fisheries bodies speak of fishery resources and maximum sustainable yield, whereas CITES speaks of threatened species and avoiding detrimental impacts on their survival. Whilst there is much scope for improvement, after a difficult start, relations between these bodies have settled down. Faced with considerable evidence that traditional fisheries management bodies were not ensuring the sustainable use of fisheries species, the increased involvement of CITES seemed inevitable. CITES' engagement was heavily backed by conservation NGOs.[172] National CITES authorities are traditionally housed within environment or forestry ministries with a terrestrial species bias, but increasingly Parties are also designating fisheries ministries in this role, which helps spread understanding and coordination. At an institutional level, CITES also lacks competence and expertise in fisheries issues, but engagement of fisheries experts is increasing. There is evidence that fishing and trade data collection on both sides was inadequate to provide a sound basis for management,[173] and competition between the organizations has led to greater collaboration on how practical difficulties can be resolved[174]

4.6   SPECIES-SPECIFIC CoP RESOLUTIONS AND DECISIONS

Aside from the listing of particular species or groups of species in the Appendices, Parties have developed a practice of adopting species-specific resolutions and decisions which do not always aim at addressing a particular problem with the implementation of CITES for the taxa concerned but rather seek to call for better implementation of existing measures for these species. The trend seems to have started around CoP9 in 1994 with resolutions on tigers (*Panthera tigris*)[175] and rhinoceros (Rhinocerotidae spp.) in Asia and Africa.[176] Subsequently, very similarly titled resolutions on bears (*Ursidae* spp.), houbara bustard (*Chlamydotis undulata*), musk deer (*Moschus* spp.), Tibetan antelope (*Pantholops hodgsonii*), Asian big cat species (Felidae spp.), sharks (Elasmobranchii spp.), great apes (Hominidae spp.), pangolins (Manidae spp.), helmeted hornbill (*Rhinoplax vigil*), snakes (Serpentes spp.), and marine turtles (Cheloniidae spp.) were agreed. Sometimes these resolutions were accompanied by suites of decisions calling for specific actions or reports. Because it is generally easier to get them adopted, the number of such decisions and of featured species has increased in recent years. Just at CoP19 in 2022, decisions were adopted for West African vultures (Accipitridae spp.), amphibians (Amphibia spp.), pangolins (*Manis* spp.), African lions (*Panthera leo*), leopards (*Panthera pardus*), songbirds (Passeriformes spp.), African grey parrots (*Psittacus erithacus*), Saiga antelopes (*Saiga* spp.), eels (*Anguilla* spp.), sharks and rays (Elasmobranchii spp.), humphead wrasses (*Cheilinus undulatus*), seahorses (*Hippocampus* spp.), queen conch (*Strombus gigas*), and marine ornamental fishes. Whole programmes of work for the Secretariat and the CITES committees have developed around these resolutions and decisions which often generate interesting information but arguably have rather nebulous objectives, except increased focus on the species concerned. As they are difficult for Parties to argue against without them appearing negative, species-specific resolutions accumulate: only one of the aforementioned resolutions has been repealed without being replaced by a new version. These species-specific activities are often driven by NGOs. They provide an excellent focus for fundraising and publicity activities, but if unchecked, they threaten to divert resources away from specific implementation problems that are a more pressing priority.

NOTES

1   Chipman, A.D. (2024) *Organismic Animal Biology: An Evolutionary Approach*. Oxford University Press, Oxford and New York.
2   Com.I.1.4.
3   Doc. 2.22.
4   Resolution Conf. 4.23.
5   Groombridge, B. (1979) *World Checklist of Endangered Amphibians and Reptiles*. Nature Conservancy Council, London. Stuart, S. (1986) *World Checklist of Threatened Birds*. Nature Conservancy Council, Peterborough. Inskipp, T.P. (1987) *World Checklist*

of Threatened Mammals. Nature Conservancy Council, Peterborough. Almada-Villela, P.C. (1988) *Checklist of Fish and Invertebrates Listed in the CITES Appendices*. Nature Conservancy Council, Peterborough.

6 Hunt, D. (1992) *CITES Cactaceae Checklist*. Royal Botanic Gardens, Kew, UK.

   Roberts, J.A., Beale, C.R. and Benseler, J.C. (1995) *CITES Orchid Checklist*, Vol. 1. Royal Botanic Gardens, Kew.

   Roberts, J.A., Allman, L.R., Beale, C.R. and Butter, R.W. (1997) *CITES Orchid Checklist*, Vol. 2. Royal Botanic Gardens, Kew.

   Davis, A.P. (1999) *CITES Bulb Checklist*. Royal Botanic Gardens, Kew.

   Roberts, J.A., Anuku, S., Burdon, J., Mathew, P., McGough, H.N. and Newman, A. (2001) *CITES Orchid Checklist*, Vol. 3. Royal Botanic Gardens, Kew.

   Newton, L.E. and Rowley, G.D. (2001) *CITES Checklist for Aloe and Pachypodium*. Royal Botanic Gardens, Kew.

   von Arx, B., Schlauer, J. and Groves, M. (2001) *CITES Carnivorous Plant Checklist*. Royal Botanic Gardens, Kew.

   Cowell, C., Williams, E., Bullough, L.-A., Grey, J., Klitgaard, B., Govaerts, R., Andriambololonera, S., Cervantes, A., Crameri, S., Lima, H.C., Lachenaud, O., Li, S.-J., Linares, J.L., Phillipson, P., Rakotonirina, N., Wilding, N., van der Burgt, X., Vatanparast, M., Barker, A., Barstow, M., Beentje, H. and Plummer, J. (2022) *CITES Dalbergia Checklist*. Royal Botanic Gardens, Kew.

7 World Conservation Monitoring Centre (1996) *Checklist of CITES Species: A Reference to the Appendices to the Convention on International Trade in Endangered Species of Wild Fauna and Flora*. CITES Secretariat/World Conservation Monitoring Centre, Lausanne, Switzerland and Cambridge.

8 https://checklist.cites.org.

9 Schouten, K. (1990) *Checklist of CITES Fauna and Flora*, A Checklist of the Animal and Plant Species Covered by the Convention on International Trade in Endangered Species of Wild Fauna and Flora.

   Schouten, K. (1992) *Checklist of CITES Fauna and Flora*, A Checklist of the Animal and Plant Species Covered by the Convention on International Trade in Endangered Species of Wild Fauna and Flora. Revised edition.

10 Resolution Conf. 9.3.

11 Anon (1976) *No. 14537 Convention on International Trade in Endangered Species of Wild Fauna and Flora (with Appendices and Final Act of 2 March 1973). Opened for Signature at Washington on 3 March 1973*. United Nations-Treaty Series Vol. 993: 243–417. https://treaties.un.org/doc/publication/unts/volume%20993/volume-993-i-14537-english.pdf (accessed 26.02.25).

12 Notification to the Parties N° 1 of 18 August 1975.

13 Heinrich, S. and Gomez, L. (2021) India's Use of CITES Appendix III. *Nature Conservation*, 44: 163–176. https://doi.org/10.3897/natureconservation.44.63688.

14 Notification to the Parties of 18 August 1995.

15 Notification to the Parties No. 2000/034 of 15 June 2000.

16 Notification to the Parties No 007 of 2 February 2007.

17 Bürgener, M. (2010) *The CITES Appendix III Listing of Abalone (Haliotis midae) Impact on Illegal Harvest and Trade, TRAFFIC Bull., 23*. TRAFFIC International, Cambridge.

18 Notification to the Parties No. 2010/010 of 25 May 2010.

19 Hinsley, A., de Boer, H.J., Fay, M.F., Gale, S.W., Gardiner, L.M., Gunasekara, R.S., Kumar, P., Masters, S., Metusala, D., Roberts, D.L., Veldman, S., Wong, S. and Phelps, J. (2017) A Review of the Trade in Orchids and Its Implications for Conservation. *Botanical Journal of the Linnean Society*, 186 (4): 435–455.

20 Doc. 1.40 (Rev.).
21 Resolution Conf. 1.1.
22 Resolution Conf. 2.17.
23 Doc. 3.30.
24 Doc. 4.36.
25 Plen. 2.12 (Rev.).
26 Resolution Conf. 3.15.
27 IUCN (2001) *18th General Assembly Resolutions and Recommendations.* Resolution 18.37 on Sea Turtle Ranching.
28 Resolution Conf. 11.16 (Rev. CoP15) and Resolution Conf. 19.5.
29 Resolution Conf. 5.21, later Resolution Conf. 7.14.
30 Resolution Conf. 8.10—later replaced by Resolution Conf. 10.14.
31 Doc. 8.50.
32 Resolution Conf. 8.20.
33 Resolution Conf. 9.24.
34 FAO (2000) *Report of the Technical Consultation on the Suitability of the CITES Criteria for Listing Commercially-Exploited Aquatic Species. Rome, Italy, 28–30 June 2000.* FAO Fisheries Report No. 629. Rome, Italy.
       FAO (2002) *Report of the second Technical Consultation on the Suitability of the CITES Criteria for Listing Commercially-exploited Aquatic Species. Windhoek, Namibia, 22–25 October 2001.* FAO Fisheries Report No. 667. Rome, Italy.
35 SC58 Inf. 6.
36 Resolution Conf. 8.3 (Rev CoP13).
37 Resolution Conf. 16.6 (Rev. CoP18).
38 Resolution Conf. 9.24 (Rev. CoP17).
39 FAO (2002) *Committee on Fisheries—Report of the Eighth Session of the Sub-Committee on Fish Trade—Bremen, Germany, 12–16 February 2002.* FAO Fisheries Report. No. 673. Rome, Italy.
40 FAO (2004) *Report of the FAO Ad Hoc Expert Advisory Panel for the Assessment of Proposals to Amend Appendices I and II of CITES Concerning Commercially-exploited Aquatic Species. Rome, 13–16 July 2004.* FAO Fisheries Report. No. 748. Rome, Italy. Rome, FAO, 51 p.
41 FAO (2025) *FAO Expert Advisory Panel for the Assessment of Proposal to Amend CITES Appendices.* Food and Agriculture Organization of the United Nations. www.fao.org/fishery/en/cites-fisheries/ExpertAdvisoryPanel (accessed 26.02.25).
42 Bräutigam, A. (Ed.) (1989) *Analyses of Proposals to Amend the CITES Appendices Submitted to the Seventh Meeting of the Conference of the Parties.* IUCN, Gland, Switzerland.
43 Universalia (2000) *Evaluation of IUCN SSC & TRAFFIC's Analyses of Proposals to Amend CITES Appendices.* Final Report. Universalia, Montréal, Canada. https://iucn.org/sites/default/files/2022-05/cites_final_report.pdf (accessed 26.02.25).
44 SC41 Summary Record.
45 Doc. 3.19.
46 Doc. 3.28.
47 Resolution Conf. 3.20.
48 Resolution Conf. 6.1.
49 Resolution Conf. 14.8.
50 International Whaling Commission (1978) *Special Meeting of the International Whaling Commission December 19 and 20 1978.* Plenary Session. https://archive.iwc.int/pages/download_progress.php?ref=420&size=&ext=pdf&k= (accessed 26.02.25).

51 Resolutions Conf. 2.7 and 2.9.
52 CoP2 Prop. 12.
53 CoP9 Prop.10.
54 Resolution Conf. 11.4.
55 See different perspectives in:
   Harland, D. (1992) *The African Elephant in International Law.* Tufts University, Medford, MA.
   Douglas-Hamilton, I. and Douglas-Hamilton, O. (1992) *Battle for the Elephants.* Doubleday, London.
   Couzens, E. (2014) *Whales and Elephants in International Conservation Law and Politics: A Comparative Study Earthscan.* Routledge, Abingdon/Oxon. and New York.
56 Com.I.1.4.
57 Doc. 1. 15.
58 Messer, K. (2000) The Poacher's Dilemma: The Economics of Poaching and Enforcement. *Endangered Species Update*, 17 (3): 50–56.
59 Douglas-Hamilton, I. (1979) *The African Elephant Action Plan. IUCN/WWF/NYZS Elephant Survey and Conservation Programme.* IUCN, Nairobi.
60 Douglas-Hamilton, I. (1980) African Elephant Ivory Trade Study: Final Report (Excerpts). *Elephant*, 1 (4): 69–99. https://doi.org/10.22237/elephant/1521731736 (accessed 26.02.25).
61 Douglas-Hamilton, I. (1980) African Elephant Ivory Trade Study: Final Report (Excerpts). *Elephant*, 1 (4): 69–99. https://doi.org/10.22237/elephant/1521731736 (accessed 26.02.25).
62 Tec. 1.3.
63 Resolution Conf. 3.12.
64 Plen. 5.5.
65 Com. II 6.4.
66 Doc. 6.21 Annex 2.
67 Doc. 7.21.
68 Doc. 7.43.8.
69 Com. I 8.9 (Rev.).
70 CoP8 Prop. 73.
71 Com. I 8.12 (Rev.).
72 Com. I 10.13 (Rev.).
73 Dec. 10.1
74 Resolution Conf. 10.10.
75 Resolution Conf. 10.10 (Rev. CoP19).
76 CoP12 Com. I Rep. 10 (Rev.).
77 SC58 Doc. 36.3 (Rev. 1).
78 CoP14 Inf. 40.
79 Dec. 14.77.
80 CoP17 Com. II Rec. 3 (Rev. 1).
81 Dec. 14.75–14.79.
82 CoP15 Inf. 68.
83 CITES Secretariat (2011) *African Elephant Fund Launched at CITES Meeting.* https://cites.org/eng/news/pr/2011/20110819_SC61.php (accessed 26.02.25).
84 SC77 Inf. 3.
85 Anon (2008) A Coalition of African States for Stronger Elephant Conservation. *CITES Afrique*, 1 (7). Species Survival Network Africa Regional Bureau.
86 Ho, I. and Lindsay, K. (2017) *Challenges to CITES Regulation of the International Trade in Live, Wild-caught African Elephants.* Amboseli Trust for Elephants/Fondation Franz Weber/Humane Society International. Annex to document SC69 Inf. 36.

87 CITES Secretariat (2019) *International Trade in Live Elephants.* https://cites.org/eng/news/statement/international_trade_in_live_elephants (accessed 26.02.25).

88 CoP17 Doc. 57.5.

89 CoP19 Doc. 66.5.

90 Meijer, W., Scheer, S., Whan, E., Wu, D., Yang, C. and Kritski, E. (2017) *Demand Under the Ban—China Ivory Consumption Research.* TRAFFIC and WWF, Beijing, China.

91 Said, M.Y., Chunge, R.N., Craig, G.C., Thouless, C.R., Barnes, R.F.W. and Dublin, H.T. (1995) *African Elephant Database 1995.* IUCN, Gland, Switzerland.

92 Said, M.Y., Chunge, R.N., Craig, G.C., Thouless, C.R., Barnes, R.F.W. and Dublin, H.T. (1995) *African Elephant Database 1995.* IUCN, Gland, Switzerland.

Barnes, R.F.W., Craig, G.C., Dublin, H.T., Overton, G., Simons, W. and Thouless, C.R. (1999) *African Elephant Database 1998.* IUCN/SSC African Elephant Specialist Group, Gland, Switzerland and Cambridge.

Blanc, J.J., Thouless, C.R., Hart, J.A., Dublin, H.T., Douglas-Hamilton, I., Craig, C.G. and Barnes, R.F.W. (2003) *African Elephant Status Report 2002: An Update from the African Elephant Database.* IUCN/SSC African Elephant Specialist Group. IUCN, Gland, Switzerland and Cambridge.

Blanc, J.J., Barnes, R.F.W., Craig, C.G., Dublin, H.T., Thouless, C.R., Douglas-Hamilton, I. and Hart, J.A. (2007) *African Elephant Status Report 2007. An Update from the African Elephant Database.* Occasional paper of the IUCN Species Survival Commission No. 33. IUCN Switzerland.

Thouless, C.R., Dublin, H.T., Blanc, J.J., Skinner, D.P., Daniel, T.E., Taylor, R.D., Maisels, F., Frederick, H.L. and Bouché, P. (2016) *African Elephant Status Report 2016: An Update from the African Elephant Database.* Occasional Paper Series of the IUCN Species Survival Commission, No. 60. IUCN, Gland, Switzerland.

93 International Union for Conservation of Nature and Natural Resources (2001) *African Elephant Species Now Endangered and Critically Endangered—IUCN Red List.* https://iucn.org/news/species/202103/african-elephant-species-now-endangered-and-critically-endangered-iucn-red-list (accessed 26.02.25).

94 Hutton, J. and Webb, G. (2002) *Legal Trade Snaps Back: Using the Experience of Crocodilians to Draw Lessons on Regulation of the Wildlife Trade.* Proceedings of the 16th Working Meeting of the IUCN-SSC Crocodile Specialist Group. Gainesville, Florida, USA, 7–10 September 2002. IUCN, Gland, Switzerland.

95 Jelden, D. (2004) *Crocodilians and the Convention on International Trade in Endangered Species of Wild Fauna and Flora (CITES).* In: Crocodiles. Proceedings of the 17th Working Meeting IUCN Crocodile Specialist Group, 66–68, Gland, Switzerland.

96 Anon (2016) *International Union for Conservation of Nature Species Survival Commission Crocodile Specialist Group.* www.iucncsg.org/pages/Farming-and-the-Crocodile-Industry.html (accessed 26.02.25).

97 Plen. 2.12 (Rev.).

98 Resolution Conf. 3.15.

99 Resolution Conf. 5.21, later Resolution Conf. 7.14.

100 Resolution Conf. 8.14.

101 Hutton, J.M. and Games, I. (Eds.) (1992) *The CITES Nile Crocodile Project.* CITES Secretariat, Lausanne, Switzerland.

102 Caldwell, J. (2024) *World Trade in Crocodilian Skins 2020–2022.* UN Environment Programme World Conservation Monitoring Centre (UNEP-WCMC), Cambridge.

103 AC22 Inf. 2 and Thorbjarnarson, J. (1999) Crocodile Tears and Skins: International Trade, Economic Constraints, and Limits to the Sustainable Use of Crocodilians. *Conservation Biology,* 13 (3): 465–470.

104 AC22 Inf. 2.

105 Crocodile Specialist Group (2011) *Crocodiles*. Proceedings of the 1st Regional Species Meeting. [IUCN] Crocodile Specialist Group, Darwin, Australia.

106 Anon (2015) *Exposed: Crocodiles and Alligators Factory-Farmed for Hermès 'Luxury' Goods*. People for the Ethical Treatment of Animals, Inc. https://investigations.peta.org/crocodile-alligator-slaughter-hermes/ (accessed 26.02.25).

107 Jorge Flores, O., MacQuarrie, K. and Portus, J. (1994) *Gold of the Andes: The Llamas, Alpacas, Vicunas and Guanacos of South America*. Patthey, Barcelona, Spain.

108 Wakild, E. (2020) Saving the Vicuña: The Political, Biophysical, and Cultural History of Wild Animal Conservation in Peru, 1964–2000. *The American Historical Review*, 125 (1): 54–88.

109 CoP2 Props 55 and 56.

110 Wakild, E. (2020) Saving the Vicuña: The Political, Biophysical, and Cultural History of Wild Animal Conservation in Peru, 1964–2000. *The American Historical Review*, 125 (1): 54–88.

111 Acebes, P., Wheeler, J., Baldo, J., Tuppia, P., Lichtenstein, G., Hoces, D. and Franklin, W.L. (2018) Vicugna Vicugna (Errata Version Published in 2019). *The IUCN Red List of Threatened Species*, 2018: e.T22956A145360542. https://doi.org/10.2305/IUCN. UK.2018–2.RLTS.T22956A145360542.en (accessed 20.10.24).

112 González, B.A., Vásquez, J.P., Gómez-Uchida, D., Cortés, J., Rivera, R., Aravena, N., Chero, A.M., Agapito, A.M., Varas, V., Wheleer, J.C., Orozco-terWengel, P. and Marín, J.C. (2019) Phylogeography and Population Genetics of Vicugna Vicugna: Evolution in the Arid Andean High Plateau. *Frontiers in Genetics*, 10: 445. https://doi.org/10.3389/fgene.2019.00445. PMID: 31244880; PMCID: PMC6562099. (accessed 20.10.24).

113 Resolutions Conf. 8.1, 11.6 and 18.8.

114 Kasterine, A. and Lichtenstein, G. (2018) *Trade in Vicuña: The Implications for Conservation and Rural Livelihoods*. International Trade Centre, Geneva, Switzerland.

115 Anon (2024) *The Brutal Cost of Quiet Luxury*. Bloomberg Originals. www.youtube.com/watch?app=desktop&v=W-SrWHEUn8o (accessed 26.02.25).

116 Cooney, R. (2019) *Harvest and Trade of Vicuña fibre in Bolivia: CITES & Livelihoods Case Study*. https://cites.org/sites/default/files/eng/prog/Livelihoods/case_studies/2.%20 Bolivia_vicuna_long_Aug2.pdf (accessed 26.02.25).

117 Anon (2025) *Loro Piana S.p.A.* https://loropiana.com (accessed 26.02.25).

118 Oldfield, S.F. (1988) *Rare Tropical Timbers*. IUCN, Gland, Switzerland and Cambridge.

119 Doc. 8.46 (Rev.) Annex 3.

120 FAO Forest Resources Division (1986) *Databook on Endangered Tree and Shrub Species and Provenances*. FAO Forestry Paper 77. Food and Agriculture Organization of the United Nations. Rome, Italy.

121 Doc. 10.52.

122 Resolution Conf. 10.13.

123 Plen. 10.7 (Rev.).

124 Buitrón, X. and Mulliken, T. (2003) *The Bigleaf Mahogany and CITES Appendix III. CITES World Issue Number 11*. CITES Secretariat, Geneva, Switzerland.

125 TRAFFIC (2002) *Appendix III Implementation for Big-leafed Mahogany Swietenia Macrophylla*. TRAFFIC International, TRAFFIC Online Report Series No. 1, Cambridge.

126 CoP12 Prop. 12.50 and Blundell, A.G. (2004) A Review of the CITES Listing of Big-Leaf Mahogany. *Oryx*, 38 (1): 84–90. https://doi.org/10.1017/S0030605304000134 Printed in the UK.

127 PC19 Summary Record.

128 CoP14 Com. I Rep. 5 (Rev. 1)/CoP14 Com. I Rep. 6 (Rev. 1).

129 Dec. 14.146.

130 CoP16 Com. I Rec. 13 (Rev. 1).

131 Resolution Conf. 14.4.

132 Decs. 9.34, 10.132 and 11.158.

133 Anon (2016) *ITTO-CITES Program for Implementing CITES Listings of Tropical Tree Species Programme newsletters.* International Tropical Timber Organization. www.itto. int/cites_programme/newsletters (accessed 26.02.25).

134 CITES Secretariat (2018) *Memorandum of Understanding between the Secretariat of the Convention on International Trade in Endangered Species of Wild Fauna and Flora (CITES Secretariat) and Secretariat of the International Tropical Timber Organization (ITTO Secretariat).* https://cites.org/sites/default/files/eng/disc/coop/MoU%20CITES-ITTO.PDF (accessed 26.02.25).

135 CITES Secretariat (2022) *CITES Tree Species Programme.* https://cites-tsp.org (accessed 26.02.25).

136 CITES Secretariat (2023) *Climate Action Through Legal and Sustainable Trade of Timber, Aromatic and Medicinal Trees: Germany Commits 10 Million Euros for the Sustainable Management of CITES Tree Species.* https://cites.org/eng/news/germany-commits-10-million-euros-to-cites-tree-species-project#:~:text=On%208%20December%202023%2C%20the,to%20the%20CITES%20Tree%20Species (accessed 26.02.25).

137 PC25 Doc. 12.

138 Reeve, R. (2015) *The Role of CITES in the Governance of Transnational Timber Trade.* Center for International Forestry Research Occasional Paper 130. Bogor, Indonesia.

139 Groves, M. and Rutherford, C. (2023) *CITES and Timber: A Guide to CITES-Listed Tree Species.* Revised edition. RutherfordGroves Publishing.

140 Doc. 11.18.

141 FAO (2004) *Report of the Expert Consultation on Legal Issues Related to CITES and Commercially-exploited Aquatic Species. Rome, 22–25 June 2004.* FAO Fisheries Report. No. 746. Rome, Italy.

142 Resolution Conf. 14.6.

143 Doc. 9.58.

144 Plen. 9.2 (Rev.).

145 Doc. 10.51.

146 Decs 10.48, 10.73, 10.74, 10.93 and 10.126.

147 FAO (1999) *Report of the FAO Technical Working Group on the Conservation and Management of Sharks. Tokyo, Japan, 23–27 April 1998.* FAO Fisheries Report. No. 583. FAO, Rome.

148 FAO (1999) *International Plan of Action for Reducing Incidental Catch of Seabirds in Longline Fisheries. International Plan of Action for the Conservation and Management of Sharks.* International Plan of Action for the management of fishing capacity. FAO, Rome, Italy.

149 Com.I 10.12 (Rev.).

150 Com. I Rep. 9 (Rev. 1).

151 Docs 10.60 and 10.60.1.

152 Com.I 10.5 (Rev.).

153 Taylor, S. (1997) The Historical Development of the Caviar Trade and the Caviar Industry. In Birstein, V.J., Bauer, A. and Kaiser-Pohlmann, A. (Eds.). *Sturgeon Stocks and Caviar Trade Workshop. Occasional Paper of the IUCN Species Survival Commission No. 17.* IUCN, Gland, Switzerland and Cambridge.

154 SC45 Summary Record.

155 Resolution Conf. 12.7.

156 AC24 Doc. 12.2.
157 Capacity building for the recovery and management of the sturgeon fisheries of the Caspian Sea (TCP/INT/3101).
158 Speer, L., Lauck, L., Pikitch, E., Boa, S., Dropkin, L. and Spruill, V. (2000) *Roe to Ruin: The Decline of Sturgeon in the Caspian Sea and the Road to Recovery.* Natural Resources Defense Council/Wildlife Conservation Society/SeaWeb.
      Harris, L. and Shiraishi, H. (2018) *Understanding the Global Caviar Market. Results of a Rapid Assessment of Trade in Sturgeon Caviar.* TRAFFIC and WWF Joint Report. TRAFFIC, Cambridge.
159 FAO (2010) *Report of the FAO and CITES Technical Workshop on Combating Illegal Sturgeon Fishing and Trade. Antalya, Turkey, 28–30 September 2009.* FAO, Rome, Italy.
160 CITES Secretariat (2023) *History of CITES Listing of Sharks (Elasmobranchii).* https://cites.org/eng/prog/shark/history.php (accessed 26.02.25).
161 CITES Secretariat (2006) *Memorandum of Understanding between the Food and Agriculture Organization of the United Nations (FAO) and the Secretariat of the Convention on International Trade in Endangered Species (CITES).* https://cites.org/sites/default/files/eng/disc/sec/FAO-CITES-e.pdf (accessed 26.02.25).
162 Nakamura, J.N. and Kuemlangan, B. (2023) *Implementing the Convention on International Trade in Endangered Species of Wild Fauna and Flora (CITES) Through National Fisheries Legal Frameworks—A Study and a Guide.* Second edition. FAO Legal Guide No. 4. FAO, Rome, Italy. https://doi.org/10.4060/cc8051en.
163 UN Trade and Development (2025) *Blue BioTrade: Promoting Sustainable Livelihoods and Conservation of Marine Biodiversity in the Caribbean Region.* https://unctad.org/project/blue-biotrade-promoting-sustainable-livelihoods-and-conservation-marine-biodiversity (accessed 26.02.25).
164 Elassar, A. (2019) Massive Tuna Nets $3.1 Million at Japan Auction. *CNN Worldwide.* https://edition.cnn.com/2019/01/05/asia/giant-tuna-sets-record-at-japan-auction/index.html (accessed 26.02.25).
165 CoP8 Props 76 and 77.
166 Doc. 8.57.
167 Hurry, G. (Ed.) (2009) *Report of the Independent Performance Review of ICCAT.* International Commission for the Conservation of Atlantic Tunas. Madrid, Spain. www.iccat.int/documents/other/perform_%20rev_tri_lingual.pdf (accessed 26.02.25).
168 CoP15 Prop. 19.
169 CoP15 Doc. 68 Annex 3.
170 Anon (2020) *Recent History of Atlantic Bluefin Tuna.* Marine Stewardship Council. www.msc.org/species/tuna/recent-history-of-bluefin-tuna (accessed 26.02.25).
171 Terje Løbach, T., Petersson, M., Haberkon, E. and Mannini, P. (2020) *Regional Fisheries Management Organizations and Advisory Bodies. Activities and Developments, 2000–2017.* FAO Fisheries and Aquaculture Technical Paper No. 651. FAO, Rome, Italy. https://doi.org/10.4060/ca7843en.
172 Vincent, A.C.J., Sadovy de Mitcheson, Y.J., Fowler, S.L. and Lieberman, S. (2014) The Role of CITES in the Conservation of Marine Fishes Subject to International Trade. *Fish and Fisheries*, 15 (4): 563–592.
173 AC33 Doc. 41 (Rev. 1) Annex 5.
174 Pavitt, A., Malsch, K., King, E., Chevalier, A., Kachelriess, D., Vannuccini, S. and Friedman, K. (2021) *CITES and the Sea: Trade in Commercially Exploited CITES-Listed Marine Species.* FAO Fisheries and Aquaculture Technical Paper No. 666. FAO, Rome, Italy. https://doi.org/10.4060/cb2971en.
175 Resolution Conf. 9.13.
176 Resolution Conf. 9.14.

5 Making the Convention Work

This chapter reviews key issues in the development of the global policy of the Convention. It focusses particularly on those issues related to obligations upon Parties that have led to compliance measures being applied.

International treaties rely on the principle of *pacta sunt servanda*—a treaty in force is binding upon the parties to it and must be performed by them in good faith.[1] It is self-evident that the ability of Parties to fully comply with the obligations of a signed treaty is fundamental to its success. Encouraging more effective implementation often requires a carrot and stick approach, but many biodiversity-related MEAs have few carrots and have struggled to find a formula to provide a stick in the form of compliance measures.

The Convention on Biological Diversity established a Subsidiary Body on Implementation in 2016, although its focus is more on reviewing and enhancing the implementation of the Convention and its protocols rather than on taking more forceful measures. The Ramsar Convention on Wetlands has no punitive sanctions for violations of or defaulting on its treaty commitments and generally relies on the goodwill of Parties and on a 'name and shame' approach.[2] The Convention Concerning the Protection of the World Cultural and Natural Heritage (World Heritage Convention) does not provide for sanctions for non-compliance by States Parties, although deletion of a qualifying site from its World Heritage List of cultural heritage and natural heritage sites of outstanding universal value is certainly a public and political embarrassment. Devoid of a compliance mechanism in its treaty, the Convention on the Conservation of Migratory Species of Wild Animals (Bonn Convention) eventually agreed by resolution in 2017 to establish one, modelled on that of CITES.[3]

As a central body for environmental interests and the provider of the secretariat for some of the biodiversity conventions, including CITES, UNEP has tried to support the development of compliance procedures, but its *magnum opus* on the subject[4] is wordy and has failed to gain much traction amongst the conventions themselves. It did however highlight the different definitions of compliance and enforcement used in MEAs. The mixing of these terms, and the confusion that it causes, has been present throughout the history of CITES. In some cases, it reflects some level of competition between those engaged in the Convention from a legal perspective and others concerned with its policing.

For the purposes of this book, compliance is defined as the fulfilment by a Party of its obligations under the Convention and related measures agreed by the Parties, whereas enforcement refers to the measures taken by Parties to detect and punish activities undertaken by individuals or companies that are in breach of the national

DOI: 10.1201/9781003542278-5

laws that the Party has put in place to comply with these treaty obligations and related measures.

Failings related to the implementation of CITES can be attributed to four broad categories: lack of political will, insufficient knowledge, lack of capacity and funding or corruption. In the early days of the Convention, when awareness of environmental issues was less of a priority for countries, lack of political will was often the principal problem. CITES has made rather ad hoc efforts to provide tools and training for Parties, but has had no dedicated financial mechanism to support Parties to improve their implementation. Corruption in CITES implementation was rather a taboo subject for many years, but it clearly exists and has recently been addressed more openly.

CITES is distinguished from most other biodiversity-related MEAs by being able to restrict or open trade, which has direct economic consequences at varying scales. It also has the ability to detect weaknesses in implementation that are significant enough to require corrective action and has the means and will to take action. It is greatly assisted by the text of the Convention with Article XIII, rather cryptically titled 'International measures', providing a basis for applying compliance measures:

1. When the Secretariat in the light of information received is satisfied that any species included in Appendix I or II is being affected adversely by trade in specimens of that species or that the provisions of the present Convention are not being effectively implemented, it shall communicate such information to the authorized Management Authority of the Party or Parties concerned.
2. When any Party receives a communication as indicated in paragraph 1 of this Article, it shall, as soon as possible, inform the Secretariat of any relevant facts insofar as its laws permit and, where appropriate, propose remedial action. Where the Party considers that an inquiry is desirable, such inquiry may be carried out by one or more persons expressly authorized by the Party.
3. The information provided by the Party or resulting from any inquiry as specified in paragraph 2 of this Article shall be reviewed by the next Conference of the Parties which may make whatever recommendations it deems appropriate.

This provision of CITES may seem modest, but it is considerably sharper than the compliance provisions in most other MEAs. It provides the springboard for expanded overall compliance procedures as determined by the Parties themselves. The development of CITES compliance procedures has been assisted by the fact that Party obligations under the Convention are specific and clear, rather than broadly worded, ideological or aspirational. It has helped foster within CITES a culture amongst the Parties to set themselves high standards.

The obligations of the treaty on Parties have often been complemented by CoP agreements, in the form of resolutions and decisions, that provide interpretation and context and, in some cases, describe measures to ensure compliance. Being a treaty concerned with trade; the Parties have been able to develop last resort sanctions for non-compliant Parties that have financial ramifications for the defaulting Party. These have taken the form of recommendations to suspend some or all CITES trade with defaulting Parties, which have proved to be a more persuasive deterrent than the simply 'name and shame' option.

In terms of identifying deficiencies in implementation of CITES that require corrective action, Article XIII.1 of the treaty refers to specific circumstances where the status of species is being adversely affected by international trade due to poor implementation, and the more general problem of ineffective implementation of the provisions of the Convention. Currently, compliance measures are applied to Parties' inadequate implementation of the following obligations:

- designating Management Authority(ies) and Scientific Authority(ies) (Article IX);
- permitting trade in CITES-listed specimens only to the extent consistent with the procedures laid down in the Convention (Articles III, IV, V, VI, VII and XV);
- taking appropriate domestic measures to enforce the provisions of the Convention and prohibit trade in violation thereof—that is, national laws and their application (Article VIII, paragraph 1);
- maintaining and submitting records of trade and periodic reports on implementation (Article VIII, paragraphs 7 and 8); and
- ensuring that a species included in Appendix I or II is not being adversely affected by trade in specimens of that species – sustainable use (Article XIII).[5]

5.1   FIRST STEPS TOWARDS A COMPLIANCE PROCEDURE

Until 2007, compliance measures were applied in a rather piecemeal fashion. Whilst the Convention became established and experience was being gained, many Parties had significant gaps in their implementation capabilities. Some joined the Convention but took little action to implement it. In the absence of any organized compliance procedures, Parties identifying any problems of compliance by other Parties were encouraged to apply the stricter domestic measures permitted under Article XIV to prohibit trade from recalcitrant Parties. Such a limited bilateral approach was however unlikely to drive improved implementation across the board, and collective action was required in the cases judged most serious.

At CoP5 in 1985 the Secretariat had drawn the attention of Parties to significant problems of CITES implementation in Bolivia,[6] and at CoP6 in 1987, its concerns were extended to Austria, France (and French Guiana), Japan, Paraguay, and the United Arab Emirates.[7] The United Arab Emirates had become a major hub for illegal trade in elephant ivory, rhinoceros horn, and snake skins amongst other products and had taken no steps to implement the Convention[8] despite being amongst the first ten States to ratify it. Repeated attempts by the Secretariat to open a dialogue with the authorities there had been rebuffed. The matter was discussed by the Standing Committee in November 1985 and, as a result, the Secretariat issued notification to Parties on 28 November 1985,[9] urging them to prohibit CITES trade with the United Arab Emirates, the first such trade suspension recommendation.

The situation in Bolivia had been a source of concern ever since the country joined the Convention in 1979. Bolivia had exported remarkable quantities of CITES wildlife products, much of it apparently sourced from neighbouring States. Some of those neighbouring States[10] submitted a proposal at CoP5 in 1985, adopted as a CoP resolution, recommending that if within 90 days Bolivia had not demonstrated to the Standing Committee that it had adopted measures to adequately implement the Convention, all Parties should refuse to accept shipments of CITES specimens accompanied by Bolivian documents, or of specimens declared as originating from Bolivia. These sanctions to remain in place until corrective measures had been put in place by Bolivia.[11] Notwithstanding the 90 day deadline decided, months of negotiations between the Secretariat, Standing Committee, and Bolivian authorities followed. Little progress was made, and 18 months later the Standing Committee finally agreed to recommend that Parties suspend any import of CITES specimens from Bolivia[12]—a decision conveyed to Parties by notification on 28 November 1986.[13] These two bold decisions, targeted at UAE and Bolivia, were a turning point in the development of CITES compliance procedures, not least because of their impact.

The United Arab Emirates announced in November 1986 that it was withdrawing from the Convention. The Secretariat feared that this was because it disagreed with the principles of CITES and wished to continue to make financial profits from the plunder of developing countries' wildlife resources.[14] However, when UAE rejoined CITES in 1990, it seemed that the withdrawal had been to give them time to try and put their house in order.

Within two months of Bolivia being formally notified about the impending trade ban, the Secretariat was invited to speak with the President of Bolivia, and a plan was agreed to remedy the situation, using funds provided by the USA.[15] The recommendation to suspend trade from Bolivia was lifted later in 1987.

Trade suspensions recommendations followed for a number of non-Parties which were engaged in undermining CITES objectives by not issuing the 'comparable documentation' required under Article X when trading in CITES specimens: Macau [Portugal] (1986),[16] El Salvador (1986),[17] the United Arab Emirates [as a non-Party] (1987),[18] and Equatorial Guinea (1988).[19] All subsequently joined (or rejoined in the case of UAE) the Convention, and it may be supposed that the action that Parties took influenced these actions.

By 1993, the Standing Committee was also flexing its muscles on some more specific issues. Concerned that measures taken by China to control illegal trade in rhinoceros horn and tiger specimens were inadequate, the Committee did not go as far as recommending a trade suspension for CITES specimens,· but reported that Parties should consider implementing stricter domestic measures up to and including prohibition of trade in wildlife species[20]

Further recommendations by the Standing Committee for Parties to suspend trade in CITES specimens due to manifest failures to implement the Convention were made in relation to Thailand (1991),[21] Italy (1992),[22] Greece (1998),[23] and the Democratic Republic of the Congo (2001).[24] Most of these were quickly lifted following action by the affected Party. After these trade suspension recommendations made in response to more generalized concerns further such recommendations were often directed at resolving more specifically defined issue. Many arose from a 'Review of alleged infractions', which was initially designed to identify attempts to violate or evade the provisions of the Convention—largely questions of enforcement—but which was broadened to include a range of other issues. A review of alleged infractions was first suggested at CoP5 in 1985 where St. Lucia, whilst not proposing the introduction of any disciplinary action for defaulting Parties, felt that a circulated written report of problems observed by the Secretariat would encourage Parties to take corrective action and to fulfil their obligations to the best of their abilities.[25] Their initiative led to agreement that at future meetings of the CoP the Secretariat should make a separate report under Article XIII of the Convention to cover all cases of alleged infractions which came to their attention. The objectives of this exercise were

a) to provide the Parties with a record of instances where it appears that significant attempts (successful or unsuccessful) have been made to violate or evade the provisions of the Convention;
b) to stimulate constructive discussion of such problems, identify those of major concern or those requiring special attention and seek mechanisms or solutions to reduce or eliminate them.

The first such 'Review of alleged infractions' was submitted to CoP6 in 1987[26] and involved 54 cases, with the diversity reported under eight categories:

A. Appendix I species traded commercially or in large quantities or without valid documents (Article III) [17 cases]
B. Appendix II species traded without valid documents (Article IV) [five cases]
C. Appendix III species traded without valid documents (Article IV) [one case]
D. Cruelty or inhumane treatment during transport (Articles III, IV and V) [two cases]
E. Failure of a party to take action against illegal trade [Article VIII, 1(a)], or to respond to the Secretariat under Article XIII [one case]
F. Use of forged or fraudulent documents [20 cases]
G. High volume trade with non-party states which undermines CITES objectives [four cases]
H. Repeated general actions of a party which diminish the effectiveness of CITES. [four cases]

Many of these cases related to single incidents or shipments. Excluded from consideration were a large number of problems associated with Party obligations such as the absence of annual trade reports. The scattergun approach provoked a very defensive reaction from many Parties, and an acrimonious debate followed with no significant outcome.[27] The Secretariat's next report on alleged infractions presented at CoP7 in 1989[28] was circulated in advance to Parties in order to allow them time to prepare comments. The diversity of infractions changed considerably, and the 81 cases featured in detail (out of 400 that the Secretariat was aware of) included infractions of

CITES obligations other than those concerning specific trade transactions in listed species. Insufficient national legislation, failure to submit annual reports, and failure to designate a Scientific Authority were all included. Two resolutions resulted from the discussions, one addressing control of trade in transit between two States,[29] and the other, confusingly titled 'Enforcement' concerning the nomenclature of parts and derivatives in trade and some general principles about the conduct of Secretariat enquiries about Article XIII designed to engender more timely responses.[30]

By the time of CoP8 in 1992, the number of categories of infractions had increased again, and the number of cases reviewed had also increased to 135. Based on a report of an in-session working group, the CoP agreed resolutions on national laws, travelling live-animal exhibitions,[31] and a compliance procedure relating to the submission of annual reports.[32] The latter was an idea that the USA, chair of the working group, had unsuccessfully proposed at CoP7 in 1989.[33] By CoP9 in 1994, the elements of a focussed suite of compliance procedures were beginning to take shape, and the Secretariat used 57 cases to illustrate a range of compliance issues which were much more closely associated with obligations of the treaty, rather than a somewhat random list of different 'categories' as in previous reviews of alleged infractions.[34] The measures consequently agreed at CoP9, however, mostly concerned improvements and clarifications of existing resolutions of the CoP, as revealed by the cases reviewed, rather than compliance procedures per se. This change reflected the fact that the title of the exercise had been changed to Review of alleged infractions 'and other problems of implementation of the Convention'. In their report to CoP10 in 1997 the Secretariat sought to define 'infraction' thus:

- illegal trade, in general with criminal intent and often without documents, or sometimes with false or falsified documents. This type of infraction is commonly committed by individuals and is similar to fraud in other fields such as drugs, weapons, etc. across the world. While restrictions exist in many areas, there will always be those who seek to profit from their abuse. Control of this abuse can only be achieved by reinforcing controls and increasing the efficiency of properly resourced enforcement officers. A particular area of concern in this area involves persons who obtain documents which appear to be valid but cover illegal specimens.
- non-compliance by Parties with the provisions of the Convention either directly or as interpreted by Resolutions. This can include the issuance and acceptance of invalid CITES documents and not implementing of basic requirements of the Convention (such as the designation of Management and Scientific Authorities or the requirement to produce annual and biennial reports). In cases such as this, the Party concerned must bear the responsibility.

The substance of the former is now more usually termed enforcement, but a Parties' inability to act on illegal trade may be considered a breach of the fundamental principle in Article III of the treaty that Parties should not allow trade in specimens of species included in Appendices except in accordance with the provisions of the Convention.

The main conclusion of the discussion of the Secretariat's document at CoP10 in 1997, reflecting concerns expressed by Switzerland and the European Union, was that a clear distinction should be made between alleged infractions of the provisions of the Convention and non-compliance with the provisions laid down in resolutions.[35] A consolidation of existing resolutions agreed at the CoP led to a different format for the review of alleged infractions and other problems of implementation of the Convention presented at CoP11 in 2000.[36] Gone was the huge list of cases from which broader lessons could be drawn. Instead, the Secretariat brought to the attention of the CoP only those instances where a Party had flagrantly or deliberately flouted or ignored the provisions of either the Convention or resolutions or where the Secretariat was unable to resolve matters of dispute—there were none of the latter. For more enforcement related matters, the Secretariat only reported on work by Parties which illustrated innovative or particularly significant enforcement action.[37]

5.2 DEVELOPMENT OF A HOLISTIC COMPLIANCE FRAMEWORK

Parties had referenced Article XIII provisions on compliance in three early resolutions,[38] but it was only at CoP11 in 2000 that more specific guidelines on its application were agreed in a resolution.[39] These guidelines stated in particular that, having been alerted by the Secretariat to a possible breach of the Convention, affected Parties were recommended to respond within one month. In the resolution, the Secretariat was instructed to bring major problems with the implementation of the Convention by a Party to the attention of the Standing Committee if they were unable to resolve them bilaterally with the Party concerned.[40]

In 2001, prompted by the issue of the late or non-submission of annual reports, the Standing Committee instructed the Secretariat to prepare an analysis of the range of legal technical and administrative actions that might be taken in response to problems of non-compliance with the Convention, resolutions, and decisions more broadly.[41] The Secretariat's thoughtful report[42] in 2002 defined a range of sequential and graduated responses to non-compliance and set the stage for a long and at times acrimonious debate on the subject in subsequent meetings of the CoP and the Standing Committee.[43] Consideration was even given to the establishment of a compliance committee,[44] but this was not agreed.

It was not until five years later, at CoP14 in 2007, that a Guide to CITES compliance procedures was agreed.[45] The sensitivity of the matter was clear from the language adopted at the CoP in relation to the Guide, with the Parties agreeing only to 'taking note' of the Guide and recommending that it just be 'referred to' when dealing with compliance matters. The Guide's opening paragraph specifically notes that it is non-legally binding, even though as an annex to a resolution, it could never be so. The Guide was however a significant milestone that has served the Parties well since. Largely echoing Article XIII.1 of the treaty, it sets out five basic obligations of the treaty to which particular attention must be paid:

i) designating Management Authority(ies) and Scientific Authority(ies) (Article IX);

ii) permitting trade in CITES-listed specimens only to the extent consistent with the procedures laid down in the Convention (Articles III, IV, V, VI, VII and XV);

iii) taking appropriate domestic measures to enforce the provisions of the Convention and prohibit trade in violation thereof (Article VIII, paragraph 1);

iv) maintaining records of trade and submitting periodic reports (Article VIII, paragraphs 7 and 8); and

v) responding as soon as possible to communications of the Secretariat related to information that a species included in Appendix I or II is being adversely affected by trade in specimens of that species or that the provisions of the Convention are not being effectively implemented (Article XIII).

Notably, failure to make agreed financial contributions to the CITES Trust Fund is not amongst the basic obligations as it is not mentioned in the text of the Convention. This contrasts with other conventions, for example, the Convention on the Conservation of Migratory Species of Wild Animals, which decided that Parties with contributions in arrears of three years or more should be denied their right to vote and to have representatives holding office in convention bodies.[46]

The initial text of the Guide to CITES compliance procedures addressed the need to take into account relevant resolutions and decisions relating to treaty obligations, with those resolutions relating to annual reports, national laws, the Review of Significant Trade (explained later in this chapter) and enforcement measures, specifically stated as being applicable. Later, the CoP-adopted measures related to the National Ivory Action Plans and the Review of Trade in Animal Specimens reported as Produced in Captivity were added to this list. Also referenced was the resolution which charges the Standing Committee with handling general and specific compliance matters, confirming it as the lead body for dealing with compliance, supported by the Secretariat. Thus, from around 2008 the Standing Committee began to take a more systematic approach to compliance matters, gathering them all in one section in their meeting agendas.

The fundamental approach in the Guide is stated as being supportive and non-adversarial in encouraging long-term compliance. However, the sequential and graduated responses first described by the Secretariat in its 2002 report can still lead to a recommendation from the Standing Committee for Parties to suspend trade in CITES specimens where a Party's compliance matter is unresolved and persistent and the Party demonstrates no intention to achieve compliance. The measures in the Guide can also be applied to non-Parties. Typically, the Secretariat draws the attention of the Standing Committee to a compliance issue within the fields identified by the CoP—in the case of non-detriment findings/biological sustainability this is under the Review of Significant Trade and normally reflects advice from the Animals and Plants Committees. The Standing Committee determines if the concerns are merited and, if so, makes recommendations to the Party concerned. The Secretariat serves as intermediary and reports on progress.

If compliance cannot be achieved, the Standing Committee can recommend to Parties to suspend commercial or all trade in specimens of one or more CITES-listed species and to inform their enforcement and Customs authorities to avoid the inadvertent acceptance of specimens of species subject to such a recommendation. Parties that require the issuance of import permits for trade in specimens of Appendix-II species are also encouraged to take account of these recommendations when processing permit applications.

After the initial rather broad use of recommendations to suspend trade because of significant weaknesses in implementation, the procedures became more targeted and rigorous, following due process as set out in the resolutions of the Parties. Although compliance policy has followed a fairly constant course in relation to national legislation and other Party obligations specifically cited in the Guide, a more unpredictable procedure for applying Article XIII emerged when considering more generalized problems in particular Parties.

Starting with Guinea in 2011,[47] the Democratic Republic of the Congo, the Lao People's Democratic Republic in 2014,[48] and Nigeria in 2018,[49] the Secretariat drew the Standing Committee's attention to more generalized or multifaceted concerns about the implementation of the Convention in certain Parties. In the case of Guinea, the issues were a lack of enforcement, possible corruption, and theft of blank permits, in later cases the concerns were much more diffuse.

For the Lao People's Democratic Republic, they were as follows: inadequate national legislation, lack of training of national CITES Authorities, weaknesses in law enforcement, poor monitoring of wildlife farms and related trade, and concerns about trade in live Asian elephants (*Elephas maximus*).[50] For the Democratic Republic of the Congo: quota management and issuance of export permits, management of exports of African grey parrots *Psittacus erithacus*, illegal trade generally, challenges in CITES implementation for African cherry (*Prunus africana*) in areas affected by human conflict, and trade in Afrormosia (*Pericopsis elata*).[51] Weaknesses attributed to Nigeria: lack of robust scientific institutions, legal optimization (grey areas or obsolete forestry policies and laws), high levels of transnational organized wildlife crime, weak national enforcement cooperation and coordination, deficient CITES controls at ports of exit, lack of capacity to fight transnational organized wildlife crime, poor handling and disposal of seized specimens, and absence of interconnected information systems.[52]

Long lists of recommendations to these Parties were proposed by the Secretariat and agreed by meetings of the Standing Committee. Often these recommendations would evolve from meeting to meeting, making it difficult for the Parties concerned to obtain a clean bill of health. This factor, coupled with the complexity of the recommendations and the requirement for regular reporting back by the Parties concerned, meant that the cases often dragged on for many years. Concerns about implementation in Guinea quickly resulted in a Standing Committee recommendation to suspend trade in 2013,[53] but in the case of the Lao People's Democratic Republic, it took nine years before such a decision was reached.[54] In the cases of the Democratic Republic of the Congo and Nigeria, the reporting and reviewing about trade in some species continues into 2025.

Meantime, the Secretariat was also initiating Article XIII compliance procedures for a rather eclectic range of other specific circumstances. Rather complex and micro-managed compliance procedures were started in relation to ebonies (*Diospyros* spp.), palisanders and rosewoods (*Dalbergia* spp.) from Madagascar, stockpiles of pangolin (*Manis* spp.) specimens in the Democratic Republic of the Congo, non-detriment findings and legal acquisition findings of 16 African Parties for African rosewood (*Pterocarpus erinaceus*), and non-detriment findings and legal acquisition findings for sharks and rays (Elasmobranchii spp.) in Ecuador.

In 2017, the Secretariat announced that it was opening an Article XIII case against Japan concerning the introduction from the sea of 90 sei whales (*Balaenoptera borealis*).[55] The reason for this sudden concern about an activity that had been going on for some years remained obscure, but it established a trend for pursuing importing Parties under CITES compliance procedures, after many years of focus on exporters from developing countries. In a similar vein, attention later turned to the registration of operations that breed Appendix-I animal species in captivity for commercial purposes by the European Union and the UK.[56] The case against Japan was resolved without resort to sanctions and those in relation to the EU and UK are ongoing in 2025.

All this places a considerable strain on the Secretariat. Whilst struggling to keep up with ongoing casework, it nevertheless continued to open new compliance cases. The Standing Committee spends an increasing amount of time on these, many of which take years to reach a conclusion.

Although the Guide to CITES compliance procedures adopted by Parties in 2007 contains a basic framework for action, other specific resolutions: on national laws, national reports, biologically unsustainable trade (known as 'Review of Significant Trade'), and on illegal trade in elephant ivory (National Ivory Action Plans), provide some level of detail on the circumstances in which compliance procedures could or should be applied.

5.2.1 INADEQUATE NATIONAL LAWS FOR THE IMPLEMENTATION OF THE CONVENTION

As CITES is not self-executing, the ability of Parties to incorporate the obligations of the Convention into national law is fundamental to good implementation. Initially driven by the IUCN Environmental Law Centre,[57] over time, guidelines were developed for national legislation which could be used to implement the Convention.[58] This culminated in a model law developed by the Secretariat in the early 1990s which has been constantly improved and updated since.[59]

Attempts to identify what laws were in place to implement the Convention within each Party were led by the IUCN Environmental Law Centre in the form of an index of species mentioned in legislation, an initial version of which was distributed at CoP2 in 1979.[60] At CoP3 in 1981, this effort was endorsed by the Parties who were encouraged to contribute to it.[61] IUCN's Environmental Law Centre continued to push matters forward. They called on the assistance of Gerhard Emonds, a CITES stalwart from the German CITES authorities, to prepare some legislative guidelines in 1981,[62] and in 1993 these were developed into a more refined tool in conjunction with the Convention's Secretariat.[63] Much later, in view of the increasing number of commercially exploited aquatic species whose trade was being regulated under the

Convention, a guide to incorporating CITES provisions into national fisheries legal frameworks was also prepared in conjunction with FAO.[64]

At CoP8 in 1992, the Secretariat concluded that for those States that had been Parties to the Convention for several years, a lack of legislation was no longer a valid excuse for poor implementation,[65] and arising from CoP discussions over alleged infractions and other problems of enforcement of the Convention, the USA drew attention to the fundamental problem of the lack of adequate national legislation for implementing CITES in many countries. A working group established on-the-spot was charged with, amongst other things, considering the role of the Secretariat in helping improve national legislation.[66] The USA chaired the working group which came up with a draft resolution on national laws for implementation of the Convention which was subsequently adopted.[67] The USA identified three key features of the resolution: that Parties lacking adequate legislation be identified by the Secretariat, the need to provide technical assistance to such Parties, and that the Secretariat should report its findings to the Standing Committee and the Conference of the Parties.[68] The Resolution remains extant in 2025 with these three features still in place. Also still present in the Resolution are the four actions which Parties should provide for in national legislation, giving them the ability to

i)   designate at least one Management Authority and one Scientific Authority [Article IX.1];
ii) prohibit trade in specimens in violation of the Convention [Article VIII.1];
iii) penalize such trade [Article VIII.1 (a)]; and
iv) confiscate specimens illegally traded or possessed [Article VIII.1 (b)]

Shortly after CoP8 in 1992, the USA went on to provide funding to the Secretariat to contract the IUCN Environmental Law Centre and TRAFFIC to undertake pilot analyses of national legislation to implement CITES. The contractors classified Parties efforts into three categories:

Category 1: legislation that is believed generally to meet the requirements for implementation of CITES.
Category 2: legislation that is believed generally not to meet all of the requirements for the implementation of CITES.
Category 3: legislation that is believed generally not to meet the requirements for the implementation of CITES.[69]

Although they refer only to whether national legislation for CITES implementation exists in a particular Party and not on whether such legislation is implemented properly, these categories have also stood the test of time as they remain the benchmarks for national legislation to implement the Convention in 2025. In subsequent years, these analyses were built upon, first to cover remaining Parties which had not been reviewed under the first contract, then to revisit Parties who had adopted new legislation, and to new Parties. Already from CoP9 in 1994, the analyses were forming the basis for pressure to be put upon Parties not having adequate legislation in place to implement the Convention. This process was led by the Standing Committee under direction from a number of CoP decisions. The Standing Committee was given the authority to recommended that all Parties suspend CITES trade with Parties whose legislation was believed generally not to meet the requirements for the implementation of CITES (Category 3) and in which significant wildlife trade occurs.[70] Thus, on 30 September 1999 following agreement in the Standing Committee, Guyana became the first Party to be subject to such a trade suspension recommendation because of inadequate legislation to implement the Convention.[71] At CoP15 in 2010, these decisions were embedded as a long-term policy in a resolution. The Standing Committee was given the authority to determine which Parties had not adopted appropriate legislative measures and to recommend appropriate compliance measures which might include a recommendation to suspend trade in CITES specimens. In parallel, efforts to support Parties in developing the necessary legislation continued. The Secretariat convened national and regional legislative workshops and recommended expert consultants to those Parties seeking to improve their own situation. However, core funding for these activities remained very limited. The whole package was branded the National Legislation Project.

By 2025, seven Parties (Djibouti, Dominica, Liberia, Libya, Oman, Sao Tome and Principe and Somalia) retain a recommendation to suspend all trade in CITES specimens as they are not believed to have legislation in place which meets the requirements for the implementation of the Convention and have not responded to offers of assistance. None of these are major wildlife trading nations. More significantly, and despite all these efforts, by 1 June 2022 only 59% of Parties were believed to have legislation in place to meet the requirements for the implementation of CITES, and a worrisome 17% (31 Parties) were believed generally not to meet the legislative requirements for the implementation of CITES.[72] The remaining 24% of Parties met some but not all of the requirements. A number of the Parties that do not have proper legislation to implement the Convention have been a Party to it for several decades. The National Legislation Project has been an innovative and thorough process, but after 20 years, this one initiative is proving insufficient to ensure that all Parties have the basic legislative tools to implement CITES.

5.2.2   Inadequate Annual Reports of Trade Authorized

The importance of annual reports detailing the extent of international trade in specimens of CITES-listed species authorized by each Party was quickly recognized. They provide a mechanism for monitoring the implementation of the Convention and the volumes of trade taking place in specimens of each species. The timing and nature of these reports was standardized with the calendar year January–December adopted as the appropriate reference period and the following 31 October as the deadline for submission. A 'uniform format' for the reports was first proposed in 1976[73] and has since been greatly expanded as 'Guidelines for the preparation and submission of CITES annual reports'.[74] The Secretariat commissioned regular analyses of the reports submitted by Parties and presented these at each meeting of the CoP.

By CoP8 in 1992, the reviews were revealing serious issues with the submission of annual reports. The Secretariat proposed, and Parties agreed, that this constituted a major problem with the implementation of the Convention and that recalcitrant Parties should be referred to the Standing Committee under Article XIII.[75] Initially such referrals resulted in naming and shaming, and there were some signs of improvement. Differences of opinion emerged in the Standing Committee over the need for recommendations to suspend trade in the event of non-compliance concerning annual report submissions, and the Committee turned to the CoP for guidance on the mandate it had to take action regarding the lack of submission of annual reports.[76] In the general consolidation of resolutions adopted at CoP9 in 1994, this question was left open.[77]

Under this more voluntary approach to annual reports, the exhortations by the Secretariat and the Chair of the Standing Committee failed to have an effect, and by CoP11 in 2000 the number of Parties submitting annual reports late, or not submitting them at all, was rising. The Secretariat's proposal that Parties not authorize any trade in specimens of CITES-listed species with a Party that the Standing Committee determined had failed to submit an annual report for three consecutive years without good excuse was agreed with little comment at CoP11 in 2000.[78] However, when it came to applying such measures, in June 2001, the Standing Committee initially deferred their application, questioning not only the efficacy of such trade suspensions but also whether they would be 'in compliance with' the World Trade Organization.[79],[80] They asked the Secretariat to produce an analysis of the range of legal, technical, and administrative actions that might be taken in response to problems of non-compliance with the Convention, resolutions and decisions more generally. The Committee relented at its meeting prior to CoP12 in 2002 and finally determined that Afghanistan, Bangladesh, Djibouti, Dominica, Liberia, Rwanda, Somalia, and Vanuatu had failed to submit the required annual reports without good excuse and thus these became the first Parties for which a recommendation to suspend trade in specimens of CITES-listed species was made for this reason.[81]

Bangladesh was subsequently found to have been included in this list in error, and within a year, all but Liberia had provided the missing annual reports. The provision for a trade suspension sanction after three consecutive years of non-submission of annual reports was changed from a short-term decision to being enshrined as a long-term resolution policy at CoP13 in 2004.

The Secretariat has continued to provide support and training, but as it has noted, the timely submission of annual reports is often down to a question of political will and administrative organization rather than technical capabilities.[82] Whilst a limited number of Parties have subsequently fallen foul of the compliance measures provided for, the sanctions agreed upon have been sufficient to ensure that the vast majority of Parties now submit their annual trade reports in a timely manner.

This is in notable contrast to the biennial report on legislative, regulatory, and administrative measures taken to enforce the provisions of the Convention required under Article VIII.7 (b). Since CoP13 in 2004, Parties have renamed these implementation reports, reduced their frequency to one per inter-CoP period (contrary to the requirements of the Convention), and provided a standard template for their production. None of these changes have halted the steady decline in their submission, which, in the absence of any compliance procedure, now stands at about 30% of Parties.[83] Whilst implementation reports are arguably of less importance to the effective implementation of the Convention, the difference between compliance rates between the two types of reports is striking.

5.2.3 PREVENTING UNSUSTAINABLE TRADE

Although charismatic endangered species listed in Appendix I often grab the media headlines, 97% of CITES-listed species are included in Appendix II. Trade in these species is permitted but regulated in order to avoid utilization incompatible with their survival in the wild and within limits that allow the maintenance of the species throughout their range at a level consistent with their role in the ecosystem in which they occur. Local people in all parts of the world often have long histories of applying effective stewardship to natural resources, which they have been using for millenia. In the international sphere, arguably starting with the UNESCO Biosphere Conference in 1968,[84] this concept has been renamed as rational or sustainable use. The concept featured strongly at the United Nations Conference on Environment and Development in Rio de Janeiro in 1992, and the sustainable use of biodiversity (including species) is one of the three principal objectives of the Convention on Biological Diversity (CBD), which entered into force in 1993. CBD defines sustainable use as follows:

> The use of components of biological diversity in a way and at a rate that does not lead to the long-term decline of biological diversity [including species], thereby maintaining its potential to meet the needs and aspirations of present and future generations.

CITES' objectives for Appendix II species is seen by some as a forerunner of this concept in international law. Its practical application is found in Article IV of the Convention where Scientific Authorities in exporting Parties are required to determine that the export of any given specimens from their country will not have a detrimental impact on the survival of the species concerned—these determinations became known as 'non-detriment findings'.

The quality of these non-detriment findings was discussed briefly at the non-plenary Special Working Session of the Conference of the Parties on implementation issues in 1977 and Sweden undertook to consider drafting a resolution on the matter, which they subsequently presented to CoP2 in 1979.[85] The resulting resolution,[86] adopted by that CoP incorporated elements related to illegal trade, but can be considered as the starting point for a compliance procedure for Parties failing to ensure that trade in Appendix II is sustainable. In keeping with the tone of Sweden's discussion document, the resolution focussed on what importing Parties could do to block trade that they considered unsustainable, calling for the use of unilateral stricter domestic measures by importing Parties if bilateral dialogue with the exporting Party (involving the Secretariat where required) was not successful—in the importing States' view.

The Resolution charged the then Committee of Technical Experts with the task of 'dealing' with control over trade in Appendix II (and III) species. In the event, it was Australia who took this matter forward at CoP3 in 1981, opining that non-detriment findings required detailed information on the total ecology of a species, which was missing in the majority of cases. Australia suggested that if importing

Parties exercised independent judgement over whether or not to allow imports, rather than only relying on export documentation, exporting Parties would be encouraged to allocate more resources to the management of Appendix II species. Further, it proposed the establishment of an expert committee to determine guidelines for criteria under which Scientific Authorities might advise Management Authorities to restrict the number of export permits for commercial trade in Appendix. II species. Not finding much support for their proposal, Australia agreed instead to act as a 'clearing house' for good practice and prepare such guidelines itself on the basis of inputs from Parties.[87]

By CoP4 in 1983, Australia had received no submissions to allow it to prepare such guidelines. Adopting a more conciliatory approach, it called for the Technical Expert Committee to identify Appendix II species subject to regular or intensive international trade and determine measures to ensure that continued trade in these species does not lead to their becoming candidates for inclusion in Appendix I. It also called for exporting and importing Parties to co-operate over actions to put these measures into effect.[88] In the resolution adopted after the discussion of their proposal, the language was softened further and the focus put on Appendix II species 'subject of significant international trade for which scientific information on the capacity of the species to withstand such levels of trade is insufficient'.[89]

The Technical Committee later formed a working group, chaired by the USA and comprised only of Switzerland, Italy, and the IUCN Conservation Monitoring Centre to review the matter. The group met over two days in December 1984 and formulated the procedure and timetable[90] for a review of significant trade in Appendix II species which is still recognizable in CITES policy today. CoP5 in 1985 adopted their protocol and identified the IUCN Conservation Monitoring Centre as the technical hub for the execution of the review. Using funds from a number of Parties, WWF and Pet Industry Joint Advisory Council, the Secretariat commissioned IUCN Conservation Monitoring Centre to conduct an initial review of Appendix II animal species of concern. Under the supervision of a working group of the Animals Committee, a group of scientists at the IUCN Conservation Monitoring Centre who were to become the engine house for the review, honed the criteria for selecting species for review, and the results were finally published in 1988 in three volumes.[91] These identified animal species for which available information indicated either over-exploitation, or for which available information was insufficient to allow a determination of the impact on wild populations of harvest for international trade. The process led not only to the identification of species being overused, but also to proposals to amend the Appendices (with both increased and decreased protection) for a number of species. The importance of the work was flagged by the chair of the Standing Committee at CoP7 in 1989:

> This work is the lifeblood of CITES. If we cannot trade Appendix II species at levels that are known to be not detrimental to their survival, then our scientific knowledge is inadequate and our treaty is threatened.[92]

Recognizing the importance of this work, despite it being a time of significant strain on the Secretariat's core budget, funds were nevertheless allocated to the continuing study of Appendix II species and to conduct some fieldwork to follow-up the findings.

The initial three-volume review of trade in specimens of Appendix II animal species taken from the wild was based on CITES-reported trade levels for animal species from 1980–1982. A second review of trade reported between 1983 and 1988 covered both animals[93] and plants.[94] Considerable effort was made to refine the criteria for identifying species in significant trade, but Parties were concerned that there was insufficient follow-up of recommendations arising from the Review.

At CoP8, Parties were faced with a series of documents tabled by Honduras, the USA, the UK, and Uruguay aiming to respond to this gap, particularly with respect to birds. After considerable debate, a compromise resolution was agreed, dealing only with animal species. It charged the Animals Committee with making recommendations for remedial action where it had identified problems with the making of non-detriment findings. It gave affected Parties a limited time to implement these recommendations, failing which, the Secretariat should recommend to the Standing Committee that all Parties immediately take strict measures, up to and including a suspension of trade in the affected species with the affected Party.[95]

The Resolution marked a bold change in efforts to address unsustainable international trade. It did however only address the results of the reviews of trade which had taken place from 1980–1982 and 1983–1988 and only applied to animal species. Based on the reports prepared by the World Conservation Monitoring Centre and IUCN, the Animals Committee immediately adopted a series of recommendations relating to 27 species, which were conveyed to the Parties concerned.[96] The recommendations were divided into two categories depending on the perceived urgency with which they needed to be implemented. At SC29 in 1993 the Secretariat reported to the Standing Committee on the implementation of these recommendations. For those that had not been implemented within the deadline, the Standing Committee, with little debate, agreed to recommend a suspension of trade in nine species from 12 Parties:

Argentina—Guanaco (*Lama guanicoe*)
Azerbaijan—Eurasian lynx (*Felis lynx*)
China—Leopard cat (*Felis bengalensis*)
Guinea—African grey parrot (*Psittacus erithacus*)
Indonesia—Lesser sulphur-crested cockatoo (*Cacatua sulphurea*)
Latvia—Eurasian lynx (*F. lynx*)
Lithuania—Eurasian lynx (*F. lynx*)
Moldova—Eurasian lynx (*F. lynx*)
Peru—Red-masked Conure (*Aratinga erythrogenys*) and grey-cheeked parakeet (*Brotogeris pyrrhopterus*)
Ukraine—Eurasian lynx (*F. lynx*)
United Republic of Tanzania—Fischer's lovebird (*Agapornis fischeri*) and pancake tortoise (*Malacochersus tornieri*)
Uzbekistan—Eurasian lynx (*F. lynx*)

These were conveyed to Parties[97] and became the first such compliance measures related to the biological sustainability of exports of Appendix II species. The implementation of these trade suspension recommendations, particularly in market Parties such as the European Union countries and the USA led them to have an immediate impact. The process gathered pace, and in 1994 a further suite of recommendations was agreed by the Animals Committee, this time relating to over 120 species and addressed to no less than 37 Parties and another set of remedial recommendations sent to these Parties in March 1996.[98] For those cases where no action was taken by the Party concerned, further Standing Committee recommendations to suspend trade in particular species from particular Parties followed. The Animals Committee also sought and obtained guidance from the CoP, in particular on objective parameters and a standard procedure for selecting candidate species for the review.[99]

The World Conservation Monitoring Centre had prepared a study on significant trade in plants species during the period 1983–1989,[100] but at CoP8 in 1992, although Parties endorsed the recommendations, they left the Plants Committee and the Secretariat to follow up its recommendations without formal process.[101] Attempts to monitor potentially detriment levels of trade being authorized by Parties for plant species were particularly hampered by poor reporting. The study revealed broader weaknesses in the implementation of the Convention for plants which were also addressed by the Plants Committee.[102] A much more detailed and bespoke programme for a review of significant trade specifically for plant species was adopted at CoP10 in 1997,[103] but by CoP11 in 2000 it was agreed that there was a need to harmonize the procedures for animal and plant species and adjust the process in the light of experience gained.[104]

Results of the Review of Significant Trade were encouraging, often manifested as either the halting or reducing of trade in species perceived as being over-exploited. The Standing Committee was able to withdraw many of its recommendations to suspend trade after affected Parties began complying with the Animals and Plants Committee recommendations. The Review of Significant Trade gathered pace, but Parties were having difficulty keeping up with the process. Multiple ongoing casework, glitches in the process discovered during implementation, overlapping and sometimes contradictory decisions on the matter taken by CoP meetings, and a tendency to over-complicate to find agreeable solutions were making the procedure burdensome and only fully understood by a limited number of insiders.

On the initiative of the Animals and Plants Committees, an evaluation of the Review of Significant Trade was mandated by CoP12 in 2002,[105] but it took two face-to-face meetings of an Advisory Working Group and 14 years to complete this exercise. The resulting revision of the resolution on the Review of Significant Trade was aimed at ensuring that the review was proportional, transparent, timely, and simple and helped build the capacity of the scientific authorities to carry out their non-detriment findings.[106] Many of these objectives were met, in particular the requirement for more dialogue to take place with Parties prior to any compliance action being taken, but simplification proved elusive.

It was not until 2023 that a Review of Significant Trade Management System[107] allowing Parties and interested observers to track the progress of species in the Review was launched with support from the United Nations International Computing Centre. In general, Parties' knowledge of the status and biology and population dynamics of their own species and the impact of international trade upon it remained patchy.

Sustainability guidance from other sources such as the Addis Ababa Principles and Guidelines for the Sustainable Use of Biodiversity[108] by CBD were reviewed by the CITES science committees but were found to be not always immediately applicable to the decision-making process under CITES, particularly with respect to making non-detriment findings.[109] Management practices applied in the field of forestry and fisheries are being introduced into CITES implementation, but the process has been slow. Efforts have been made to provide Parties with tools to assist the making of non-detriment findings[110] and to promulgate examples of good practice.[111] Two major CITES workshops have been held to review and improve guidance materials—in Cancun, Mexico, in 2008[112] and Nairobi, Kenya, in 2023.[113] In general, however, support for Scientific Authorities to undertake their duties under the Convention remains scarce, both in terms of knowledge and resources.

The Review of Significant Trade has encouraged Parties to move in the direction of more biologically sustainable trade in specimens of species listed in Appendix II. Whilst in many cases the consultations with affected Parties by the Animals, Plants, and the Standing Committees have resulted in positive actions to reduce or halt trade until more is known about its impact on wild populations of the species, this is not always the case. The Standing Committee has felt compelled to issue recommendations to suspend trade in certain species from certain countries on many dozens of occasions since the first instances of such recommendations in 1993. A procedure set out in the Review of Significant Trade resolution[114] calls for a review of such recommendations to suspend trade that have been in place for longer than two years to evaluate the reasons why this is the case in consultation with the range State and, if appropriate, take measures to address the situation, but even this has not managed to unblock the situation in some cases. Some trade suspension recommendations for species from Madagascar and the Solomon Islands have now remained in place for almost 30 years with no resolution in sight. Whilst this may help solve the immediate problem, it does not fit well with the principle of the Guide to CITES compliance procedures for such measures to be supportive and non-adversarial with the aim of ensuring long-term compliance. Whilst there can be little doubt that trade in specimens of Appendix II species has been made more sustainable by the Review of Significant Trade, too often this has been achieved by bringing the trade concerned to a complete halt, either voluntarily or by outside pressure. Avoiding biological extinction of species caused by international trade has proved achievable, but reaching consensus on levels of use that are ongoing and biologically sustainable remains rather elusive.

5.2.4 Enforcement

Article XIII of the Convention provides a basis for compliance measures to be applied where Parties have done too little to enforce the provisions of the Convention and prohibit illegal trade. However, the subject of enforcement has had a curious history in the development of the Convention. The word 'enforcement', although originating from old French, is more often associated with English-speaking countries, particularly the USA and UK than other cultures. Indeed, in CITES usage, there is no direct translation of the word in French and Spanish, but it is expressed as 'application of

the Convention'. Early references to enforcement in CITES often referred both to preventative action to avoid breaches of the Convention by Parties, as well as the penal
action against individuals and companies with which the word is now more often
associated.[115]

Enforcing laws on international trade in wildlife was not always a priority for
national police and inspection authorities or international bodies with a mandate
covering law enforcement. Coinciding with the adoption of the Convention in
1973, an Indonesian initiative at the International Criminal Police Organization
(INTERPOL) led to a survey of its national central bureaux to see which countries
were importing and exporting wild animals and how any control measures were
applied and enforced. The response was modest and showed that the main responsibility for protecting wild animals and controlling their import and export rested with
agencies other than the police. It was concluded that INTERPOL involvement with
the matter could only have a marginal effect. The INTERPOL General Assembly
did pass a resolution urging national central bureaux to take action within their
power on cases of traffic in wildlife which was illegal in the country of origin and to
encourage co-operation on this matter.[116] In 1974 INTERPOL offered its cooperation
for the enforcement of the Convention once it came into force[117]

From the start, the Secretariat followed up cases of poor implementation, including cases of illegal trade that it became aware of or were drawn to its attention by
NGOs such as TRAFFIC. However, its resources were too limited to take much
action. The Parties complained that they did not have the capability to investigate all
the cases drawn to their attention either.[118] As early as CoP2 in 1979, some Parties
were calling for the establishment of a committee of technical experts on enforcement of wildlife laws to analyze and discuss patterns and trends in illegal trade
and to review and develop enforcement measures to alleviate illegal trade.[119] The
Secretariat advised that the establishment of another CITES committee was to be
avoided and Parties instead requested the Secretariat to maintain its cooperation
with INTERPOL.[120] At the same CoP, the Secretariat submitted a list of cases where
it had detected possible enforcement problems and presented these in the context
of Article XIII of the Convention.[121] Details of each case were brief, but even this
effort established that the inventory was too burdensome to detail, especially as the
number of Parties increased.

At CoP3 in 1981, the newly established Technical Expert Committee was charged
with identifying problems with enforcement of the Convention and providing guidance to the Secretariat and the Parties on measures to remedy them.[122] With an
already large mandate, the Committee did not address this specific point but instead
focussed on interpretation and technical issues that were hampering enforcement
efforts. Identification of specimens to species and the question of what parts and
derivates are covered by the Convention were two major foundation issues in this
respect.

As the Secretariat observed at CoP4 in 1983, without proper enforcement, the
Convention is a totally useless piece of paper.[123] The idea of a regular review of
enforcement efforts was not resurrected until the Secretariat's 'Review of Infractions'.
This began at CoP6 in 1987 with the objective of reviewing instances where attempts
had been made to violate or evade the provisions of the Convention and seeking

solutions to reduce or eliminate them. Early lists of infractions presented by the Secretariat to CoP under this review included many instances of poor enforcement of the Convention by Parties, but in time the Review was extended to cover other weaknesses in implementation and it lost its enforcement focus. The Review of Infractions highlighted particularly flagrant cases of weak enforcement usually addressed by the Party concerned, and a more concerted effort to raise enforcement standards was missing. In CITES discussions in the late 1980s there was considerable overlap between the concept of enforcement and implementation more broadly. A resolution adopted at CoP7 in 1989 titled 'Enforcement' was largely restricted to descriptions of parts and derivatives with some guidance about compliance procedures generally.[124]

The late 1980s and early 1990s saw a flurry of other actions that were to change the direction and pace of CITES enforcement activities. At international level, in 1988, the Customs Co-operation Council published an enforcement training module devoted to CITES and later established a CITES Enforcement Working Group. For their part, INTERPOL published a practical guide for CITES Management Authorities on how to cooperate with INTERPOL Central Bureaux[125] and provided a special CITES supplement to its magazine *International Police Review*. These actions immediately led to a large increase in activities on wildlife trade within INTERPOL, and in 1994 its Environmental Crime Working Group created a subgroup on wildlife crime. In the USA, a Wildlife Forensic Laboratory was opened in 1988 which over time greatly improved the capabilities of policing and prosecuting authorities, not only in the US, but globally.

National wildlife trade enforcement authorities were also starting to work more closely together. In Africa, a first African Wildlife Law Enforcement Co-operation Conference in Zambia in December 1992 led to the establishment, under the auspices of UNEP, of an intergovernmental agreement known as the Lusaka Agreement on Co-operative Enforcement Operations Directed at Illegal Trade in Wild Fauna and Flora (Lusaka Agreement), which came into effect in 1996. In North America, an American Wildlife Enforcement Group was created in 1994 to formalize the exchange of intelligence and training related to wildlife regulations enforcement between Canada, Mexico, and the USA.[126] The member States of the European Union also established a CITES Enforcement Group in legislation which came into force in 1996.[127]

A number of mostly anglophone Parties began a concerted push within CITES to establish some sort of enforcement working group. In the words of the USA, one of the proponents, the subject was

> the interdiction of illegal flora and fauna in international trade; the investigation and subsequent prosecution of illegal wildlife (flora and fauna) traders; and the gathering and dissemination of information that will assist Parties in the detection of illegal wildlife trade and the interdiction of illegal CITES specimens.[128]

This parallels the definition now more widely used for the term enforcement.

At CoP7 in 1989 Canada called for the establishment of a CITES Enforcement Committee,[129] but its focus and objectives needed more thought, and further consideration of the modalities was deferred.[130] Although Papua New Guinea raised the issue again at CoP8 in 1992, the Secretariat noted that the Standing Committee had already rejected the idea of establishing an Enforcement Committee.[131] Later, the Animals Committee suggested a Law Enforcement Network be established, and the Standing Committee asked the Secretariat to consult Parties by Notification about the idea.[132] The response was limited. Although Australia, South Africa, Switzerland, the UK, and the USA supported the proposal, others were opposed citing confidentiality concerns in particular, and the Standing Committee decided to defer consideration of the issue again until CoP9 in 1994.

At CoP9, Ghana proposed increased focus on enforcement by Parties and the Secretariat, including the establishment of a Law Enforcement Consultative Group under the Standing Committee.[133] In a vote, 22 Parties were in favour of the creation of such a consultative group, but 50 Parties were opposed. The remaining text of the adopted resolution (minus the consultative group proposal) did however turn a spotlight on the need for better enforcement and addressed the lack of capacity in the Secretariat for this activity.[134] The USA was not finished with the idea of an enforcement working group, and at CoP10 in 1997 they proposed the establishment of a Working Group on Illegal Trade in CITES Specimens. After a rather acrimonious debate, the matter was again put to a vote, with 35 Parties in favour and 45 against.[135] In 2004, the CoP requested a meeting of experts to assist the coordination of investigations regarding violations of the Convention and to help maintain appropriate levels of confidentiality regarding law enforcement information.[136] The participants at the resulting meeting, hosted by the USA, restyled themselves as the CITES Enforcement Expert Group and issued an extensive statement.[137] The group met again in the USA in 2009 and reported back to the Standing Committee.[138] Amongst its conclusions was that the lack of data about the scale and nature of illegal trade in wildlife was hampering the efforts to develop enforcement strategies and responses at the national, regional, and international levels.

Although there were some existing data-gathering efforts in this regard, such as the Wildlife Enforcement Monitoring System (WEMS) Initiative by the United Nations University and an alliance of NGOs,[139] Parties preferred to develop something themselves and asked the Secretariat to establish an illegal trade database working group, to design and implement a database to gather and analyze data related to illegal trade in specimens of CITES-listed species.[140] Parties were subsequently requested to submit information for this database on administrative measures (e.g., fines, bans, suspensions) imposed for CITES-related violations; significant seizures, confiscations, and forfeitures of CITES specimens; criminal prosecutions or other court actions; and disposal of confiscated specimens during the year 2013.[141] The establishment of a central database proved burdensome for the Secretariat, but at CoP17 in 2016 a formal reporting obligation on Parties to provide a CITES illegal trade report annually in line with a format agreed by the Standing Committee was agreed.[142] It is still in force today, but failure to submit such reports is not however subject to any compliance procedures.

Following a request from the UN General Assembly to collect information on patterns and flows of illicit trafficking in wildlife and to report thereon,[143] The United Nations Office on Drugs and Crime (UNODC) had already begun to use data submitted by CITES Parties to produce a global overview report on illegal trade and, in 2016, produced the first of what became a quadrennial World Wildlife Crime Report.[144] Parties had been reluctant to support the cost of producing such a report and directed the Secretariat to share their data with ICCWC to assist global research and analysis studies. The Secretariat had some concerns over the use of Parties' data by an outside body and at first tried to exercise some control over the narrative used in the UNODC reports by providing extensive editorial comments. However, lack of staffing resources made this too burdensome in the longer term.

Building on the pioneering work of the Lusaka Agreement and networks in Europe and North America, a series of similar Wildlife Enforcement Networks (WENs) started being established. Initially these were supported technically by the Secretariat, but later ICCWC took on this role, and global meeting of Wildlife Enforcement Networks (WENs and similar bodies) have been held in the margins of CITES CoPs since CoP16 in 2013. To date, seven WENs have been created:

- Association of Southeast Asian Nations Working Group on CITES and Wildlife Enforcement (ASEAN-WEN)—established 2005
- South Asia Wildlife Enforcement Network (SAWEN)—established 2008
- Red de Observancia y Aplicación de la Normativa Silvestre para Centroamérica y República Dominicana (Central America WEN)—established 2010
- Central Africa Wildlife Enforcement Network (Central Africa WEN)—established 2012
- South America Wildlife Enforcement Network (SudWEN)—established 2014
- Caribbean Wildlife Enforcement Network (CaribWEN)—established 2016
- Horn of Africa Wildlife Enforcement Network (HA-WEN)—established 2017

Others are under consideration in different parts of the world. Often initiated and funded by outside donors including NGOs, they are sometimes established under intergovernmental processes, in other instances they are more informal alliances. They provide a regional or subregional focus for information sharing, coordinated action, and other activities to combat illicit wildlife trade.

At meetings of the CoP, the focus on combating illegal trade in specimens of CITES-listed species was increasingly included in species-specific resolutions and decisions about high-profile species such as elephants, bears, and tigers. Rather than creating a new dedicated enforcement working group within CITES, attention focussed on improving coordination with existing bodies with expertise in this field at both international and national levels. The Secretariat signed a memorandum of understanding with the World Customs Organization in 1996, with INTERPOL and with the US Fish and Wildlife Service Office of Law Enforcement/Clark R. Bavin National Fish and Wildlife Forensic laboratory in 1998 and with the Lusaka Agreement Task Force in 2000.

The Secretariat had always been rather lukewarm about the suggestion of establishing an enforcement working group. Part of the reason was their concern about their lack of capacity to service such a body. In response, in 1990, the USA seconded

John Gavitt, a former Special Agent with the Fish and Wildlife Service, to the Secretariat 'to reinforce its potential to fight against infraction'.[145] He was the first in a series of secondments of experts with a police or law enforcement background to the Secretariat, most of whom came from the USA or the UK. These staff permitted the Secretariat to initiate a number of initiatives on enforcement issues. In 2000, a computerized system (TIGERS—Trade Infraction and Global Enforcement Recording System) to process reports of wildlife crime and illicit trade from CITES Management Authorities or enforcement agencies and a variety of sources was established[146] only to be discontinued in 2007 when maintaining it became too labour-intensive and the process became too haphazard and incomplete to reach its potential.[147] This was followed by the communication of alerts about current trends in wildlife offences or information about new modi operandi related to the illegal trade, sent via the restricted access enforcement authorities' forum on the CITES website and also sent to WCO and INTERPOL for onward communication through their respective networks.[148] This too fell into disuse and in 2015 was incorporated into a similar communication tool used by WCO.[149] In the military/policing tradition, from 2002 the Secretariat started awarding Secretary-General's Certificates of Commendation for exemplary enforcement actions—a similar Secretary-General's Certificate of Merit for excellence in other areas of CITES implementation was begun in 2011, but has not persisted.

As part of a broader consolidation of existing resolutions, the rather dispersed references to enforcement in CoP resolutions were brought together in one place at CoP11 in 2000.[150] The review of alleged infractions begun in 1987 was wound down, and instead the Secretariat reported regularly on work by Parties which illustrated innovative or particularly significant enforcement actions.[151] This was the start of a pattern of reporting by the Secretariat to the CoP which kept enforcement in the spotlight.

To facilitate cooperation, in 2004, Parties had been recommended to convey the contact details of their relevant national law enforcement agencies responsible for investigating illegal trafficking in wild fauna and flora to the Secretariat. However, progress on this was slow and despite energetic follow-up, by 2010 less than half the Parties had done so. Whilst recognizing the role that NGOs can play in the enforcement of the Convention, the Secretariat was concerned that in some cases, government agencies appeared to have almost abdicated their statutory and constitutional role in this regard to NGOs, who appeared to be gaining access to enforcement-related information in a manner that it considered legally questionable.[152] The Secretariat announced its intention to meet representatives of INTERPOL, UNODC and WCO in late 2009 to adopt a common enforcement strategy that could be followed internationally but which would also benefit national activities.

The Secretariat subsequently chaired a meeting of INTERPOL, UNODC, the World Bank and WCO in Vienna, Austria, in November 2009. After some high-level prodding by incoming Secretary-General John Scanlon, a Letter of Understanding between these organizations establishing the International Consortium on Combating Wildlife Crime (ICCWC) was signed at an International Tiger Forum in Saint Petersburg, Russian Federation, in November 2010. The news was announced to Parties with surprisingly little fanfare at a meeting of the Standing Committee in

2011.[153] It proved to be the crowning achievement of former British policeman John Sellar, who was initially seconded to the Secretariat by the UK in 1997 and later became the first established staff member with a law enforcement background. In 14 years, he did much to raise the profile of enforcement within CITES and increase awareness of the importance of wildlife trade issues amongst partner enforcement organizations who often had other pressing priorities.

Notwithstanding the sometimes-muted engagement of CITES Parties, combatting illegal trafficking in wildlife moved rapidly up the agenda in the wider international sphere. In 2014, the UN Security Council adopted two resolutions in the context of its sanctions regime, that explicitly recognized that illicit exploitation of wildlife and wildlife products was facilitating support for armed groups and criminal networks in the Democratic Republic of the Congo and Central African Republic and thus exacerbating conflicts in the region.[154] The next year, the UN General Assembly adopted its first dedicated resolution on tackling illicit trafficking in wildlife recognizing the legal framework provided by and the important role of CITES, and the important work of ICCWC.[155] Also in 2015, the UN Sustainable Development Summit included the following targets addressed at tackling illegal trafficking in its 2015–2023 Sustainable Development Goals:

- Take urgent action to end poaching and trafficking of protected species of flora and fauna and address both demand and supply of illegal wildlife products.
- Enhance global support for efforts to combat poaching and trafficking of protected species, including by increasing the capacity of local communities to pursue sustainable livelihood opportunities.[156]

The UN Economic and Social Council's Commission on Crime Prevention and Criminal Justice had adopted a resolution on illicit trafficking in protected species of wild flora and fauna in 2001.[157] Other forums such as the G20[158] and the United Nations Environment Assembly[159] passed resolutions on illegal trade in wildlife and momentum grew. Money flooded into ICCWC and combatting illegal trade more generally on the back of this high-level support. The Global Environment Facility agreed a USD 90 million global wildlife programme with a strong emphasis on reducing the impacts of illegal trafficking on protected species. According to the World Bank, between 2010 and 2018 in excess of USD 2.3 billion was spent on combatting wildlife trafficking globally.[160] Some of this would concern CITES-listed species. In order to combat illegal trade in wildlife at least, Parties now have considerable carrots to counterbalance the modest sticks that CITES possesses.

The establishment of ICCWC was a key event leading to this increased focus. Although the day-to-day running of ICCWC was overseen by the Secretariat through its chairing of ICCWC's Senior Experts Group, the full mission of the Consortium to strengthen criminal justice systems and provide coordinated support at national, regional, and international level to combat wildlife and forest crime exceeded CITES' competence and mandate. ICCWC quickly developed a life of its own with a vision statement, five-year strategic work programme, website and annual reports. Its budget surpassed that of CITES many times over. The activities of ICCWC are still reported by the Secretariat at each meeting of the Standing Committee and CoP but are often noted with little debate or discussion. It supports some activities identified as important by CITES Parties, but its actions have to be agreed by all ICCWC partners, and some of them sometimes have different priorities than the CITES Parties. With so much funding available, the scramble for influence and position in the field of combatting illegal trade in wildlife amongst international bodies is often intense. Curiously, after leading the creation and institutionalization of ICCWC, today CITES sometimes appears as a bystander.

As in other circumstances with significant financial implications, corruption is regrettably a feature of the regulation of wildlife trade. The Convention was however rather slow to address this matter specifically at policy level. In 2015, on the occasion of the 6th Session of the Conference of States Parties to the UN Convention against Corruption, the then Secretary-General and the Executive Director of UNODC issued a joint statement drawing attention to its importance in the context of wildlife trade.[161] The following year, the Secretariat raised the matter in the Standing Committee.[162] The Committee agreed that the issue of corruption is increasingly relevant and of great importance. As a result it submitted a draft decision to CoP17 in 2016 requesting ICCWC to develop guidelines that could be used to promote adequate integrity policies and assist Parties to mitigate the risks of corruption in the trade chain as it relates to CITES-listed specimens. This was subsequently adopted.[163] At the same CoP, the European Union and Senegal proposed a resolution on prohibiting, preventing, detecting, and countering corruption which facilitates activities conducted in violation of the Convention.[164] It called for a much greater focus on this issue by Parties. UNODC subsequently produced guidance for wildlife management authorities on this matter,[165] and it has become a core feature of ICCWC training and capacity-building. At CoP19 in 2022, Parties agreed to encourage the development of national corruption risk mitigation policies and strategies to address corruption risks associated with wildlife crime, so that swift and decisive action can be taken where corrupt activities are detected.[166]

Despite the huge developments in CITES policy on illegal wildlife trade and the engagement of partner organizations, prioritization of activities and recognition of the enforcement weaknesses that may merit CITES compliance measures remains weak and rather ad hoc. Enforcement deficiencies are sometimes cited in more general compliance reviews about poor implementation of the Convention in certain Parties, but so far only Nigeria has been subject to a Standing Committee recommendation to suspend CITES trade solely due to insufficient enforcement of the Convention.[167]

5.2.5 Preventing Illegal Trade in Ivory

The decision of CoP12 in 2002 to permit a second sale of legal ivory stocks by Botswana, Namibia, and South Africa was accompanied by an instruction to the Secretariat to assess whether Cameroon, China, the Democratic Republic of the Congo, Djibouti, Ethiopia, Japan, Nigeria, Thailand, Uganda, and the USA, as countries with active internal ivory markets, had adequate legislative, regulatory, and enforcement measures in place to control internal trade in ivory. Those that did not were to submit an action plan with timelines to do so.[168] The Standing Committee

was given the authority to recommend restrictions on the commercial trade in specimens of CITES-listed species to or from non-compliant Parties.[169] With the exception of Djibouti, all the Parties concerned undertook some activities, and Cameroon,
the Democratic Republic of the Congo, and Nigeria developed action plans.[170] At the
subsequent CoP13 in 2004, the focus had subtly turned to African elephant range
States, who, as part of a general Action Plan for the Control of Trade in African
Elephant Ivory, were required to

a) prohibit the unregulated domestic sale of ivory (raw, semi-worked or
   worked). Legislation should include a provision which places the onus of
   proof of lawful possession upon any person found in possession of ivory
   in circumstances from which it can reasonably be inferred that such possession was for the purpose of unauthorized transfer, sale, offer for sale,
   exchange or export or any person transporting ivory for such purposes;
b) issue instructions to all law enforcement and border control agencies to
   enforce existing or new legislation rigorously; and
c) engage in public awareness campaigns publicizing existing or new prohibi-
   tions on ivory sales.[171]

In cases where Parties had not implemented these measures or in relation to Parties
'where ivory is found to be illegally sold', it was, rather surprisingly, the Secretariat
which was given a full mandate to recommend to Parties that they should suspend
commercial trade in specimens of CITES-listed species.[172] Normally such a decision-making role is reserved for the Parties, acting through the Standing Committee.
In consequence perhaps, although implementation was patchy, the Secretariat did
not initially find the need to recommend any trade suspensions. It did however warn
two Parties that it might do so, resulting in a positive response. In particular the
Secretariat did not believe that it would be appropriate to issue recommendations
for suspension of trade based solely on a failure to submit a progress report about
measures taken to comply with the Action Plan for the Control of Trade in African
Elephant Ivory.[173]

The Action Plan for the Control of Trade in African Elephant Ivory was renewed
at CoP14 in 2007, but after pressure from African elephant range States (in particular
Kenya and Mali),[174] it was extended again to apply to non-African elephant range
States with an ivory carving industry or internal trade in ivory that was unregulated. The compliance element of the Action Plan was also sharpened. Using the
report of the Elephant Trade Information System (ETIS) to CoP14 as an evidence
base, Parties affected by illicit trade in ivory were required to submit responses to a
questionnaire about their domestic ivory trade controls by 31 December 2007. In the
absence of a reply, the CoP agreed that all Parties were to be recommended not to
authorize commercial trade in any specimens of CITES-listed species until the completed questionnaire was returned. No less than 82 Parties[175] were required to return
the questionnaire. The Secretariat informed Parties that completed questionnaires
had not been received from Chad, the Democratic Republic of the Congo, Ethiopia,
Gabon, Guinea, Mozambique, Nepal, Nigeria, Rwanda, Somalia, Sri Lanka, Sudan,
and Swaziland and consequently recommended that Parties should not authorize

commercial trade in specimens of CITES-listed species with these countries.[176] Most of these Parties responded rapidly, and the recommendations were lifted. However, Gabon and Somalia remained under a trade suspension for four years for failing to return the questionnaire.

Returning a questionnaire was a first step, and the Action Plan for the Control of Trade in African Elephant Ivory was revised at CoP15 in 2010 to focus more on implementation, particularly in Parties that the Secretariat believed had active and unregulated internal markets for ivory or were to be significantly affected by illicit trade in ivory. Where the Secretariat found the country not implementing the Action Plan, or where significant quantities of ivory were found to be illegally sold, after consulting the Standing Committee, it was authorized to issue a Notification to the Parties advising that the CoP recommends that all Parties suspend commercial trade in specimens of CITES-listed species with the Parties in question.[177] The Secretariat admitted that it did not have the resources to support Parties in their implementation of the Action Plan or even to visit them to monitor its implementation.[178] Therefore in 2012, the Standing Committee agreed on a change of tack. Any Party identified in the ETIS analysis as being involved in substantial illegal ivory trade as a source, transit, or destination country was requested to submit a written report on its ivory trade controls by 1 January 2013.[179] As a result, nine Parties were required to develop 'National Ivory Action Plans'[180] [NIAPs]. At the subsequent CoP, the ETIS analysis proposed the categorization of such Parties into those of 'primary concern', 'secondary concern', and of 'importance to watch', and in 2014 the Standing Committee subsequently required 11 Parties of 'secondary concern' or 'importance to watch' to prepare NIAPs.[181]

In the face of evidence of increased illegal trade in ivory and powered by pressure from NGO groups and an enthusiastic Secretariat, the Standing Committee agreed that compliance warning letters be sent to Parties not supplying NIAPs. The absence of response from some Parties meant that trade suspension recommendations relating to all commercial trade in specimens of CITES-listed species soon followed. The first in 2015 were to the Democratic Republic of the Congo, Lao People's Democratic Republic, and Nigeria,[182] followed by Angola in 2016.[183] The trade suspensions were for failing to submit a plan or, in the case of Angola, a progress report—a sanction which the Secretariat had previously said would be inappropriate in such circumstances.[184]

In a retrospective attempt to codify an increasingly unstructured set of activities, CoP17 in 2016 adopted guidelines for the application of the NIAPs process and included them in a resolution on trade in elephant specimens.[185] The rather byzantine procedures still apply today.[186] Belatedly, at CoP18 in 2019, the Guide to CITES compliance procedures was amended[187] to include only the NIAPs part of the resolution on trade in elephant specimens as one of the resolutions whose poor implementation could lead to compliance procedures. The NIAPs though often require actions from Parties which are in excess of the obligations of the resolution.

Given the fairly fundamental issues to which CITES compliance measures are normally applied, those related to illegal trade in ivory might seem rather trivial, but they should be seen as a reaction of their time. Even though the peak of illegal elephant killing for international trade may have already passed when the compliance measures related to this matter started, there was huge pressure on CITES at the time to be seen to be doing something about illegal ivory trade. The rather disjointed action plans and other initiatives which were developed in response did result in more documentation and discussion, but the impact on illegal killing and trade of ivory is unclear. The punitive compliance measures adopted so far relate only to failure to supply reports and not to the efficacy of any action—that is largely determined by self-assessments made by affected countries.

5.2.6 Captive Bred and Artificially Propagated Specimens

When the Convention began, breeding of rare wild animals in captivity was relatively small-scale and often associated just with zoos. Artificial propagation of plants such as orchids and cacti was however more frequent, if not on the commercial scale that it is today. Provision was therefore made in the text of the Convention for reduced controls on specimens that had been bred in captivity or artificially propagated. Under Article VII, paragraphs 4 and 5 of the Convention, specimens of Appendix I species bred in captivity or artificially propagated for commercial purposes are treated as if they were specimens of species in Appendix II. Appendix I species bred in captivity or artificially propagated for non-commercial purposes and specimens of Appendix II species bred in captivity or artificially propagated are to be issued with a certificate confirming that fact, rather than a regular CITES permit or certificate. This appears fairly straightforward, but in fact has proved rather contentious and has led to lengthy debate, resulting in implementation guidance which are extremely complex and difficult for Parties to apply.

Difficulties have been experienced in both describing when a specimen can be described as bred in captivity or artificially propagated and under what circumstances this can be considered as being done for commercial versus non-commercial purposes. The definitions of 'bred in captivity' and 'artificially propagated' are currently spread over four different resolutions adopted by Parties. Summary guidance to help Parties determine when a specimen can be considered bred in captivity or artificially propagated is currently being revised to accommodate further changes in definitions related to artificial propagation. However, examination of the decision flow charts for issuing CITES documentation for these species in the current version of the guidance[188] is quite enough to demonstrate the nature of the problem. The applicable definitions of bred in captivity and artificially propagated have evolved over many years and been amended piecemeal with no overall strategy.

The Parties' treatment of specimens that have been bred in captivity or artificially propagated is handicapped by a lack of agreement about whether such production is broadly a positive action for the conservation of the species concerned or not. At the time of the entry into force of the Convention, the underlying policy assumption was that production in captivity was likely to affect the survival of the species less than the removal of animals from the wild.[189] An early resolution urged all Parties to encourage captive breeding of animals for pets.[190] Later, Parties also recognized that artificial propagation of specimens of plant species included in Appendix I reduces the collecting pressure on wild populations and, thus, has a positive effect on their conservation status.[191] This was in line with wider policy agreements such as the

United Nations Conference on Environment and Development in Rio de Janeiro, Brazil, in 1992. This agreed that governments should encourage and support the husbandry and cultivation of wild species for improved rural income and employment, ensuring economic and social benefits without harmful ecological impacts.[192] On the other hand, at CoP14 in 2007, Parties agreed that tigers *Panthera tigris* should not be bred in captivity for trade in their parts and derivatives. Some CITES observer organizations are opposed to trade in animals in particular, bred in captivity or not. Evidently, the acceptability of trade in animal specimens bred in captivity is not universally agreed and may be species-dependant.

A large majority of reported trade in CITES-listed plants (81% by weight and 96% by number of individuals)[193] relates to specimens that have been artificially propagated. The scale is impressive. Between 1996 and 2015 more than one billion artificially propagated orchid specimens were recorded in the CITES trade database.[194] In recent years, the proportion of specimens of CITES-listed animal species in trade that is from captive sources has also increased considerably, with over half of all reported commercial trade in CITES animal species involving specimens from non-wild sources.[195] Overall, just 18% of reported CITES trade now relates to wild-sourced individuals.[196] Scholars have identified a number of hypotheses to explain the shift in reported volumes of wildlife trade from wild-sourced to captive-sourced products. These include market demands—reliability of supplies, quality and nature of products, and increasingly negative public perception of wild-sourced products; decline in quantities of specimens available in the wild and the effects of CITES limitations on their trade; and misreporting of products sourced from wild specimens (either by accident or on purpose).[197]

In 2011, the European Union and its Member States and the USA drew the attention of the Standing Committee to their concern that trade in wild-taken animal specimens of CITES-listed species, declared as bred in captivity (or ranched), was occurring globally on a large scale, which may be having a detrimental impact on populations of the species concerned in the wild.[198] By declaring the specimens as bred in captivity, CITES permit applicants were taking advantage of the weaker controls which apply to such specimens. Unlike other exemptions and derogations to the basic provisions applying to trade in specimens of CITES-listed species, the guidance developed by the Parties for trade in specimens bred in captivity or artificially propagated has not been clear or collected in a single document and has therefore been subject to different interpretations by Parties. The Secretariat was asked to organize a number of reviews of the matter,[199] but these did not result in much greater clarity in overall policy or reduce its complexity. In an attempt to explain what provisions should apply in a given circumstance, guidance for inspecting captive breeding and ranching facilities,[200] and for deciding what controls to apply and when[201] were developed in 2017. These were followed by the development of a mobile and tablet application tool (CapaCITES) to assist with their application in 2022.[202]

The initiative by the European Union and the USA did eventually lead the Standing Committee to propose, and CoP16 in 2013 to adopt, a decision calling for the consideration by the Animals and Standing Committees of the need for a process for reviewing the implementation of CITES concerning trade in specimens that are claimed to be produced via captive breeding or ranching.[203] The Animals Committee concluded in 2014 that a procedure was required to address the incorrect use of derogations for specimens claimed to have been bred in captivity. Also for such cases to be investigated and timely measures taken to ensure compliance with the Convention.[204] In an unusual move, the Committee went on to draft a fully-fledged resolution describing a compliance procedure for these circumstances.[205] Although this was only labelled a provisional version,[206] the text was forwarded with little debate *via* the Standing Committee[207] to CoP17 in 2016 where it was adopted.[208] Reflecting perhaps a lack of preparation and political support, its implementation was subject to external resources being available. The procedure was modelled on the Review of Significant Trade with identification of problematic trade by the Animals Committee, dialogue with the Parties concerned, and recommendations for remedial measures, backed up by a trade suspension recommendation from the Standing Committee if these were not met.

In the first round of reviews under this process, between CoP17 and CoP18 and made possible by funding from the European Union and the USA, 23 species being exported from 15 Parties were identified as of concern. Following explanations from the Parties involved, ten species from 12 different Parties were subject to compliance recommendations. The procedure experienced some delays caused by the COVID-19 pandemic, but through dialogue, solutions to bring the Party into line with the Conventions provisions in relation to the trade concerned were found for all but three cases by 2023.[209] Consequently no trade suspension recommendations have been required to date. This first round of the review proved long and laborious, covering trade permitted between 2011 and 2015. A further phase of the review was initiated after CoP19 in 2022, using funds provided by Switzerland to examine trade authorized by Parties between 2017 and 2021. The Animals Committee selected 24 species from 16 different Parties for detailed review.[210] At CoP18 in 2019, Parties added the resolution on trade in animal specimens reported as produced in captivity to the list of those for which the Standing Committee may decide to recommend a suspension of commercial trade in CITES-listed specimens in the event of non-compliance.

Whilst this formal process was developed and implemented, Parties also developed parallel initiatives to investigate other trade involving specimens being afforded trade exemptions due to the fact that the specimens concerned were claimed to be bred in captivity. Included amongst these were trade in Asian big cats in variety of Parties,[211] Asian elephants being sent from Lao People's Democratic Republic to China,[212] and bird and reptile species in the European Union[213] and the UK.[214] It seemed that the procedure described in the resolution on reviewing trade in animal specimens reported as produced in captivity was not only rather slow, but was failing to pick up all the cases of concern. In retrospect, the resolution may have been developed rather too quickly and without sufficient preparation. Given that it has oversight over well over half of all commercial trade in specimens of CITES-listed animal species, its procedures allow detailed review of only a handful of cases each triennial. These are based on trade transactions that took place several years previously and may not represent contemporary trade in the species from the country concerned.

5.3   CITES COMPLIANCE MEASURES IN SUMMARY

The CITES compliance procedures are widely viewed as groundbreaking for a multilateral environmental agreement. Parties have identified the key provisions of the Convention to which they should be applied, adopted an overall procedure in 2007, and separate processes for each key provision. Having done so, they take their obligations under the Convention seriously and have applied the compliance measures with some vigour. Nonetheless, this jumble of procedures is quite complex to understand and administer. Practitioners are aware of these subtleties, but it may not be so straightforward for those not regularly involved with CITES. Although the Standing Committee as the lead body for the application of the compliance measures has tried to favour dialogue to ensure compliance and Parties go to considerable lengths to avoid being targeted, the list of those Parties that have been affected by recommendations to suspend trade in specimens of CITES-listed species has become long (Table 5.1). No less than 82 Parties, 44% of all Parties, have been subject to a recommendation to suspend trade under the CITES compliance measures.

TABLE 5.1

Recommendations to Suspend Trade in CITES Specimens Agreed under CITES Compliance Procedures

	Reasons for trade suspensions				
Country affected	Generalized problems of implementation	National legislation	Annual reports	Control of ivory trade	Unsustainable trade in Appendix II species
Afghanistan			2002–2003, 2013–date		
Algeria		2004–2005			
Angola				2016–2016	
Argentina					√
Azerbaijan					√
Bahrain					√
Bangladesh	2023–present (birds only)		2002–2003		
Belize					√
Benin					√
Bolivia	1985–1987				
Cameroon	2022–present*				
Central African Republic	2022–present*		2004		
Chad	2022–present*		2022–2022	2008–2008	
China					√
Comoros		2025–present			√
Cote D'Ivoire					√

TABLE 5.1 (*Continued*)

Recommendations to Suspend Trade in CITES Specimens Agreed under CITES Compliance Procedures

	Reasons for trade suspensions				
Country affected	Generalized problems of implementation	National legislation	Annual reports	Control of ivory trade	Unsustainable trade in Appendix II species
Democratic Republic of Congo	2001–2002 2016–present**** 2018–present*****			2008–2008, 2015–2015	√
Djibouti		2004–date	2002–2003, 2013–2013, 2018–present		
Dominica			2002–2003, 2018–2018, 2024–present		
Ecuador	2024–present******				
El Salvador	1986–1987				
Equatorial Guinea	1988–1992	2004–2004			√
Ethiopia				2008–2008	
Fiji		2002–2003			√
Gabon				2008–2012	
Gambia	2022–present*	2004–2005	2014–2014		√
Ghana					√
Greece	1998–1999				
Grenada			2016–present		√
Guinea	2013–present 2023–present*			2008–2008	√
Guinea-Bissau	2022–present*	2004–2008, 2016–2018	2004–2006		√
Guyana		1999–1999			
Haiti					√
Iceland			2019–2019		
India		2004–2005			√
Islamic Republic of Iran					√
Italy	1992–1995				
Kazakhstan					√
Lao PDR	2017–2018/2018–present*** 2023–present			2015–2015, 2016–2016	√
Latvia					√

(Continued)

TABLE 5.1 (*Continued*)

Recommendations to Suspend Trade in CITES Specimens Agreed under CITES Compliance Procedures

	Reasons for trade suspensions				
Country affected	Generalized problems of implementation	National legislation	Annual reports	Control of ivory trade	Unsustainable trade in Appendix II species
Lebanon		2025–present			
Lesotho			2013–2018		
Liberia		2004–2008, 2016–present	2002–2005		
Libya		2024–present	2011–2012, 2022–present		
Lithuania					√
Madagascar	2016–present**	2010–2014			√
Malawi					√
Mali	2022–present*				√
Mauritania		2004–2019	2003–2010		
Moldova					√
Mongolia		2025–present			
Mozambique		2004–2004			√
Myanmar					√
Nepal				2008–2008	
Nicaragua					√
Niger					√
Nigeria	2005–2011 [enforcement] 2018–present*	2008–2008		2008–2008 2015–2016	√
Oman			2024–present		
Panama		2004–2004			
Peru					√
Russian Federation					√
Rwanda		2004–2010	2002–2003	2008–2008	√
Saint Vincent and the Grenadines			2018–2018		
Sao Tome and Principe		2022–present	2016–2017, 2022–2025		
Senegal		1999–2000			√
Sierra Leone		2004–2004			
Solomon Islands			2016–2016, 2019–2019		√
Somalia		2004–date	2002–2003, 2006–2016, 2024–2025	2008–2012	

TABLE 5.1 (*Continued*)
Recommendations to Suspend Trade in CITES Specimens Agreed under CITES Compliance Procedures

	Reasons for trade suspensions				
Country affected	Generalized problems of implementation	National legislation	Annual reports	Control of ivory trade	Unsustainable trade in Appendix II species
South Sudan					√
Sri Lanka				2008–2008	
Sudan				2008–2008	√
Suriname					√
Swaziland				2008–2008	
Thailand	1991–1992				
Togo	2022–present*				√
Ukraine					√
United Arab Emirates	1985–1990, 2001–2002				
United Republic of Tanzania					√
Uzbekistan					√
Vanuatu			2002–2003		
Vietnam		2002			
Yemen		2002			

   * Non-detriment findings and legal acquisition findings for African rosewood *Pterocarpus erinaceus*
  ** Non-detriment findings and legal acquisition findings for Malagasy ebonies *Diospyros spp.* and palisanders and rosewoods *Dalbergia spp.*.
 *** Non-detriment findings and legal acquisition findings for rosewoods *Dalbergia* spp.
**** Non-detriment findings for African grey parrot *Psittacus erithacus*
***** Stockpiles of pangolins *Manis* spp.
****** Non-detriment findings and legal acquisition findings for sharks and rays *Elasmobranchii* spp.

5.4   OTHER DEVELOPMENTS WITHOUT COMPLIANCE MEASURES

5.4.1   CAPACITY-BUILDING AND TRAINING

Whilst addressing poor implementation of the Convention by Parties proved popular and quite effective, building capacity in Parties to deliver the Convention's objectives has proved more challenging. In the early days of the Convention, it was often a case of learning by doing, and establishing good practice simply took time. For many customs and border inspection staff, the arrival of a convention about wild animals and plants to enforce was a complete novelty.

Some Parties, such as Canada, went about training their port of entry staff even before the Convention came into force.[215] Early efforts by the Secretariat focused on bilateral training of national CITES officials in the course of missions to various Parties, but by 1980, using donor funding, the Secretariat hired consultants to visit Parties and provide more systematic training to customs officials in particular.[216] A German-funded seminar for enforcement officials from 22 African and Asian States in 1980 was amongst the earliest examples of international cooperation for this purpose.[217] At CoP3 in 1981 a resolution was passed on technical cooperation recognizing the difficulties being experienced by developing country Parties in training management authorities and calling on all Parties to contribute to the efforts to remedy this problem.[218] Thereafter the Secretariat organized a series of training workshops, mostly using external donor funding. At CoP7 in 1989, the Secretariat announced its intention to develop a training programme that contributes to a better understanding of the Convention and its implementation,[219] but it was not until the late 1990s that more significant developments took place. In 1997, an external study on the effectiveness of the Convention highlighted the need for developing countries and countries with economies in transition to train their personnel and equip them with appropriate facilities.

In the same year, the Secretariat established a Capacity Building Unit whose brief included training and assisting Parties to improve their national capacities to implement the Convention. The Secretariat went on to agree a Memorandum of Understanding about capacity building with the NGO TRAFFIC in 1999. It designated the local offices of TRAFFIC around the world as 'CITES Capacity Building Collaborating Centres' charged with leading capacity building activities on a regional and subregional level,[220] but this never took off as envisaged. Yet awareness of the importance of the issue continued to develop amongst Parties. In 1997, Spain announced that it was organizing a postgraduate course on conservation and international trade focussed on CITES at the International University of Andalusia.[221] The first CITES Master's course on Management, Conservation, and Control of Species Subject to International Trade was subsequently held in 1998, under the leadership of the indefatigable Margarita Clemente Muñoz, long-time chair of the Plants Committee. The course is still running today. It has been supported by the Secretariat and is now offered in Spanish, English, and French. Since its establishment it has been completed by a remarkable 418 students from 107 different Parties. A significant number of graduates have taken time out from a national role in CITES implementation to undertake the course, and between them they have established a cohort of graduates, many dozens of whom attend each meeting of the CoP. The importance of this training and the contacts and networking it engenders cannot be overstated.

In specialist fields like enforcement, partner organizations such as the World Customs Organization have also established training initiatives with the support of the Secretariat. Some of this has been undertaken under the umbrella of the Green Customs Initiative[222] established in 2004 to enhance the capacity of customs and border control officers to enforce and foster compliance with trade-related environmental conventions. Latterly, other enforcement-related training has been coordinated through ICCWC.[223] Parties also provide CITES training and capacity-building for developing countries bilaterally. Although welcome, these efforts are often not well coordinated with institutional efforts by international actors. These actions are frequently delivered under contract by NGOs who sometimes lack sufficient expertise and impartiality to do them well.

In 2010, the Secretariat received a donation of USD 3.5 million from the European Commission for a multi-year project to strengthen the capacity of developing countries to implement the Convention. This enabled a large number of activities to be undertaken including the launching of a CITES Virtual College in 2011 to provide a range of online tools and resources for Parties and others. After its initial development, it proved difficult to find the time and resources to keep the materials in the Virtual College up-to-date, but it was relaunched with fresh funding in 2022.[224] Whilst the funding from the European Commission was welcomed, some Parties found that work undertaken was not well adapted to their technological, logistical, and equipment needs.[225] At CoP16 in 2013, the Secretariat was instructed to issue a questionnaire to understand these needs better and to review the Convention's capacity-building activities to assess whether and how they could be rationalized and consolidated. This eventually led to the Secretariat producing a rather modest capacity-building framework in 2018.[226]

Some discontent with the Secretariat's efforts remained, with the USA proposing a draft resolution at CoP18 in 2019 to respond to the 'urgent need for greater coordination, transparency, and accountability in capacity-building efforts, and for meaningful assistance for Parties facing challenges'.[227] After lengthy intersessional discussions, this finally led to the adoption by the Parties at CoP19 in 2022 of a consolidated resolution on capacity-building[228] with the capacity-building activities of Secretariat under closer supervision and the Standing Committee instructed to develop an integrated capacity-building framework. To date, this has not been progressed due to lack of funding.[229]

5.4.2 Harnessing Technology

One of the most striking changes in society since the advent of the Convention has been the development of information and communication technology. Even if public bodies such as CITES, and the national authorities responsible for its application, are not generally known for rapidly adapting to change, developments in the fields of communications and information management have revolutionized the implementation of the Convention.

Although now themselves outdated, the Secretariat acquired its first telex machine in 1987 and first fax machine in 1988, both were to greatly speed up the communication with the Parties and others. The CITES website, now a vital tool for communication, was developed by UNEP-WCMC and was first registered in August 1997. It now receives over 1 million unique users per year with nearly 4 million pageviews. Social media followed, with CITES Instagram, X (formerly Twitter), LinkedIn, Facebook, and YouTube accounts created by the Secretariat.

The first meeting of the Technical Expert Committee on Harmonization of Permit Forms and Procedures in 1980 noted that computerization of the CITES permitting process could help standardize data collection and facilitate the processing of data for annual reports and their eventual comparative tabulation. At the time, some Parties were said to be 'in the process of computerization for the purpose of the Convention'.[230] This discussion led to a recommendation at CoP3 in 1981 that Parties studying or developing computerization programmes for licensing and reporting trade under the Convention consult with each other, and with the Secretariat, in order to exchange information on the computer language used and to ensure optimal harmonization and compatibility of systems.[231] At CoP5 in 1985, the Parties noted that the use of computers can help to provide trade statistics more effectively and urged all Parties to consider whether the preparation of their statistical reports could be computerized or undertaken under a contract between the Party and WTMU[232] (later UNEP-WCMC). UNEP-WCMC and its predecessors which had been contracted by the Secretariat to collect and store data from the Parties' annual trade reports was particularly advanced in the use of computers at that time and played a large part in gearing up Parties to use this new technology in the early years.

At CoP6 in 1987, the company that had equipped the UK CITES Management Authorities with a computer system gave a demonstration and the Secretariat worked with them to ensure that basic CITES requirements were met by the system and that it was compatible with UNEP-WCMC's database.[233] Initially, computerized annual report data was transferred to UNEP-WCMC by physical diskette, but reports for 1992 from Brazil and the USA were transferred directly from their Management Authorities to UNEP-WCMC—a first for this type of communication.[234] By CoP10 in 2000, 26 Parties were submitting their annual reports electronically on computer tapes, diskettes, or by electronic mail.[235]

In March 1994, the Secretariat initiated a study of the needs of the Parties for a standardized system for the production of annual reports and the management of CITES-related information on computer. A Notification on this subject was sent to the Parties with a questionnaire.[236] This led to the development of an information management strategy for the Convention in 1997,[237] which was rolled out as funds became available.[238] The theme was also taken up in the Convention's Strategic Vision 2000–2005,[239] whose action plan contained objectives of a rather general nature calling for the development and use of appropriate technologies and information management systems to enhance the collection, submission, and exchange of accurate information.

At CoP13 in 2004, Ireland (on behalf of the European Union) proposed a more concerted effort to guide the use of electronic CITES permitting,[240] but many Parties felt that developing detailed guidelines was premature in the absence of funding[241] and the Secretariat did not believe it had either the expertise or resources to undertake the preparation of guidance relating to computerized systems. The arrival of Marcos Silva, a new Secretariat staff member with experience in information and communication technology, proved timely and after first being requested by Parties at CoP14 in 2007,[242] a CITES electronic permitting toolkit was launched at CoP15 in 2010.[243] Revised versions were produced in 2013 and 2022. The principal aim of the toolkit was to harmonize CITES electronic permitting with international data standards for e-trade and protocols for electronic data exchange. There was a particular need to do this for systems developed by the United Nations Centre for Trade Facilitation and Electronic Business, WCO, the International Air Transport

Association, and the Automated System for Customs Data developed by the United Nations Conference on Trade and Development.

Within CITES, oversight for the process of keeping CITES guidance in harmony with wider international developments has been provided by a Standing Committee working group, first established in 2015 and which has operated continuously since under various names, largely chaired by Switzerland. The working group, in conjunction with the Secretariat, has developed a number of other tools under the banner of the eCITES Implementation Framework[244] including guidelines and specifications for electronic permit information exchange of CITES permits and certificates.[245] Although guidance is in place, by 2023 it was reported that only 31 Parties had made significant progress in implementing eCITES systems (with an additional 12 Parties in the early planning stage).[246] Due to the fast-moving nature of the subject, many of the developments on information and communication technology in CITES in recent years have been referenced only in a series of technical decisions adopted by the meetings of the CoP since 2004 and have probably not received the visibility or recognition that they merit.

The organization of CITES meetings has also benefitted from new technology. Since 1997, paper documents have largely been dispensed with allowing participants more time to study documents circulated electronically or via the CITES website. The Proceedings of the meetings of the CoP were produced in paper form up to and including CoP11 in 2000 and thereafter were promulgated electronically. Holding meetings online was not contemplated until Parties were compelled to do so by the COVID-19 pandemic. The first meeting of the Standing Committee online was in May 2021, with the application of the Rules of Procedure adapted for this purpose.[247] It had a deliberately limited agenda, avoiding contentious issues such as compliance, but it went remarkably smoothly and attracted 826 participants, including many from Parties not normally able to participate in its physical meetings for financial reasons. The following meetings of the Animals and Plants Committees in the May and June of the same year adopted the same approach but with a much more ambitious agendas and achieved satisfactory results. Parties though have been lukewarm about extending the use of online meetings. In part this may be due to the enhanced opportunities for dialogue at face-to-face meetings, but it is doubtless also for delegates' personal reasons—international travel often being seen as a perk of the job. The COVID-19 pandemic resulted in a perhaps overdue examination of contingency plans in the event of exceptional circumstances necessitating postponement of a CITES meeting, with online meetings being one of the options available.[248]

After a slow start, the process of using information and communication technology to modernize the implementation of CITES is now moving at some pace.

NOTES

1 Vienna Convention on the Law of Treaties (1969) Article 26.
2 Ramsar (2016) *An Introduction to the Ramsar Convention on Wetlands*. Seventh edition. Ramsar Convention Secretariat, Gland, Switzerland. www.ramsar.org/sites/default/files/documents/library/handbook1_5ed_introductiontoconvention_final_e.pdf (accessed 25.02.25).

3 UNEP/CMS/Resolution 12.9.
4 Bruch, C. and Mrema, E. (2006) *Manual on Compliance with and Enforcement of Multilateral Environmental Agreements.* UNEP, Nairobi, Kenya.
5 Resolution Conf. 14.3 (Rev. CoP19).
6 Doc. 5.46.
7 Doc. 6.19.1 and Doc.6.20.
8 Doc. 6.19 (Rev.).
9 Notification No. 366 of 28 November 1985.
10 Doc. 5.46.
11 Resolution Conf. 5.2.
12 SC14 Summary Record.
13 Notification No. 413 of 28 November 1986.
14 Doc. 6.20 (Rev.).
15 Doc. 6.20 (Rev.).
16 Notification No. 371 of 16 January 1986.
17 Notification No. 398 of 4 July 1986.
18 Notification No. 438 of 31 March 1987.
19 Notification No. 494 of 5 September 1988.
20 Notification to the Parties No. 774 of 15 October 1993.
21 Notification to the Parties No. 636 of 22 April 1991.
22 Notification to the Parties No. 675 of 30 June 1992.
23 Notification to the Parties No. 1998/35 of 6 August 1998.
24 Notification to the Parties No. 2001/039 of 9 July 2001.
25 Doc. 5.33.
26 Doc. 6.19 (Rev.).
27 Plen. 6.6 and 6.7.
28 Doc. 7.20.
29 Resolution Conf. 7.4.
30 Resolution Conf. 7.5.
31 Resolution Conf. 8.16.
32 Resolution Conf. 8.7.
33 Doc. 7.20.2.
34 Doc. 9.22 (Rev.).
35 Dec. 10.122.
36 Doc. 11.20.1.
37 Doc. 11.20.1.
38 Resolution Conf. 2.6, Resolution Conf. 3.6 and Resolution Conf. 3.9.
39 Resolution Conf 11.3.
40 Resolution Conf. 11.3.
41 SC45 Summary Record.
42 SC46 Doc. 11.3.
43 Reeve, R. (2004) *The CITES Treaty and Compliance: Progress or Jeopardy*? Chatham House Sustainable Development Programme Briefing Paper 04/01, Royal Institute of International Affairs, London.
44 SC54 Inf. 3.
45 Resolution Conf. 14.3.
46 UNEP/CMS/Resolution 9.14.
47 SC61 Doc. 30.
48 SC65 Summary Record.
49 SC70 Summary Record.

50 SC67 Doc. 12.1.
51 SC66 Summary Record.
52 SC70 Doc. 27.3.5.
53 Notification to the Parties No. 2013/017 of 16 May 2013.
54 Notification to the Parties No. 2023/127 of 21 November 2023.
55 SC69 Doc. 29.1 (Rev. 2).
56 SC74 Doc. 28.1.
57 Doc. 3.20.
58 Emonds, G. (1981) *Guidelines for National Implementation of the Convention on International Trade in Endangered Species of Wild Fauna and Flora.* IUCN, Gland, Switzerland.
de Klemm, C. (1993) *Guidelines for Legislation to Implement CITES.* IUCN, Gland, Switzerland and Cambridge.
Shine, C. and de Klemm, C. (1999) *Guidelines for Legislation to Implement CITES.* Second edition. IUCN, Gland, Switzerland and Cambridge. Unpublished.
59 CITES Secretariat (2021) *Model Law on International Trade in Wild Fauna and Flora.* Revised draft. https://cites.org/sites/default/files/projects/NLP/E-Model_law-revised_Oct.2021.FINAL.DRAFT.pdf (accessed 25.02.25).
60 Doc. 2.29 Annex 1.
61 Resolution Conf. 3.19.
62 Emonds, G. (1981) *Guidelines for National Implementation of the Convention on International Trade in Endangered Species of Wild Fauna and Flora.* IUCN, Gland, Switzerland.
63 de Klemm, C. (1993) *Guidelines for Legislation to Implement CITES.* IUCN, Gland, Switzerland and Cambridge and Shine, C. and de Klemm, C. (1999) *Guidelines for Legislation to Implement CITES.* Second edition. IUCN, Gland, Switzerland and Cambridge. Unpublished.
64 Nakamura, J.N. and Kuemlangan, B. (2020) *Implementing the Convention on International Trade in Endangered Species of Wild Fauna and Flora (CITES) Through National Fisheries Legal Frameworks: A Study and a Guide.* Legal Guide No. 4. FAO, Rome, Italy. https://doi.org/10.4060/cb1906en.
65 Doc. 8.19 (Rev.).
66 Com.II 8.2.
67 Resolution Conf. 8.4.
68 Com.II 8.11.
69 Doc. 9.24 (Rev.).
70 Dec. 10.18.
71 Notification to the Parties No. 1999/65.
72 CoP19 Doc. 28.
73 Notification to the Parties No. 25 of 1 June 1976.
74 Annex 1 of Notification to the Parties 2023/132 of 24 November 2023.
75 Resolution Conf. 8.7.
76 SC31 Summary Record.
77 Resolution Conf. 9.4.
78 Dec. 11.37.
79 SC45 Summary Record.
80 This contention was later addressed by the Secretariat: WTO and CITES Secretariats (2015) *CITES and the WTO Enhancing Cooperation for Sustainable Development.* WTO and CITES Secretariats.
81 Notification to the Parties No. 2002/064 of 19 December 2002.
82 CoP12 Doc. 22.1.

83 CITES Secretariat (2024) *Implementation Report.* https://cites.org/eng/resources/reports/Implementation_report (accessed 25.02.25).
84 UNESCO (1968) *Intergovernmental Conference of Experts on the Scientific Basis for Rational Use and Conservation of the Resources of the Biosphere. Paris, France, September 4–13, 1968.* United Nations Educational, Scientific, and Cultural Organization, Paris, France.
85 Doc. 2.16.
86 Resolution Conf. 2.6.
87 Plen. 3.9 (Rev.).
88 Doc. 4.20.
89 Resolution Conf. 4.7.
90 CoP 5 Doc. 5.26.
91 Broad, S., Luxmoore, R. and Jenkins, M. (Eds.) (1988) *Significant Trade in Wildlife: A Review of Selected Species in CITES Appendix II. Volume 1 Mammals.* International Union for Conservation of Nature and Natural Resources/Secretariat of the Convention on International Trade in Endangered Species of Wild Fauna and Flora.
    Luxmoore, R., Groombridge, B. and Broad, S. (Eds.) (1988) *Significant Trade in Wildlife: A Review of Selected Species in CITES Appendix II. Volume 2 Reptiles and Invertebrates.* International Union for Conservation of Nature and Natural Resources/Secretariat of the Convention on International Trade in Endangered Species of Wild Fauna and Flora; and Inskipp, T., Broad, S. and Luxmoore, R. (Eds.) (1988) *Significant Trade in Wildlife: A Review of Selected Species in CITES Appendix II. Volume 3 Birds.* International Union for Conservation of Nature and Natural Resources/Secretariat of the Convention on International Trade in Endangered Species of Wild Fauna and Flora.
92 CITES (1991) *Proceedings of the Seventh Meeting of the Conference of the Parties. Lausanne, Switzerland, 9 to 20 October 1989.* Secretariat of the Convention, Lausanne, Switzerland, p. 20.
93 Doc. 8.30.
94 Doc. 8.31.
95 Resolution Conf. 8.9.
96 Doc. 9.33.
97 Notification to the Parties No. 737 of 20 April 1993.
98 Doc. 10.55.
99 CoP9 Decision 1 directed to the Animals Committee and Decs 10.79 10.80, 10.81, 10.82, 11.93, 11.95, 11.106, 11.107, 11.108 and 11.117.
100 Doc. 8.31.
101 Plen. 8.8 (Rev.).
102 Doc. 9.34 (Rev.).
103 Doc. 10.56.
104 Resolution Conf. 8.9 (Rev.).
105 Decs 12.75 and 13.67.
106 AC27/PC21 Doc. 12.1.
107 CITES Secretariat (2023) *Review of Significant Trade Management System.* https://rst.cites.org/public (accessed 25.02.25).
108 Secretariat of the Convention on Biological Diversity (2004) *Addis Ababa Principles and Guidelines for the Sustainable Use of Biodiversity (CBD Guidelines) Secretariat of the Convention on Biological Diversity.* Montreal, Canada.
109 CoP14 Doc. 13.
110 CoP11 Inf. 11.3.

111 CITES Secretariat (2025) *CITES Virtual College: Non-Detriment Findings.* https://cites.org/eng/virtual-college/ndf (accessed 25.02.25).

112 Anon (2008) *International Expert Workshop on CITES Non-Detriment Findings Cancun, Mexico, November 17th to 22nd, 2008.* www.conabio.gob.mx/institucion/cooperacion_internacional/TallerNDF/wfunctioning.html (accessed 25.02.25).

113 CITES Secretariat (2023) *Workshop on Non-Detriment Findings (NDFs).* https://cites.org/eng/node/138336 (accessed 25.02.25).

114 Resolution Conf. 12.8 (Rev. CoP18), paragraph 1 (p).

115 Doc. 7.7.

116 Anon (1976) *Police Intervention and Co-Operation in the Traffic in Wild Animals in Anon (1976) 45th Session of the ICPO-Interpol General Assembly, Accra, Ghana, 14–20 October 1976.* www.interpol.int/content/download/18149/file/1976%20-%2045th%20GA%20Accra.pdf (accessed 25.02.25).

117 Doc. 2.10.

118 Doc. 3. 6.

119 Com 2.4.

120 Resolution Conf. 2.6.

121 Doc. 2.6.

122 Resolution Conf. 3.5.

123 Doc. 4.8.

124 Resolution Conf. 7.5.

125 Notification to the Parties No. 508 of 25 November 1988.

126 Anon (2010) *Strengthening Wildlife Enforcement.* Commission for Environmental Cooperation. www.cec.org/strengthening-wildlife-enforcement-1/ (accessed 25.02.25).

127 European Union (1996) Council Regulation (EC) No 338/97 of 9 December 1996 on the Protection of Species of Wild Fauna and Flora by Regulating Trade Therein. Article 14.3.

128 Doc. 10.29.

129 Com. II 7.8.

130 Com. 7.15.

131 Com. II 8.12 (Rev.).

132 Notification No. 776 of 23 November 1993.

133 Doc. 9.25.1.

134 Resolution Conf. 9.8.

135 Com. II 10.3 (Rev.).

136 Dec. 12.88.

137 CoP13 Doc. 23 Annex 1.

138 SC58 Doc. 23 Addendum.

139 Anon (2020) *Wildlife Enforcement Monitoring System.* United Nations University. https://wems-initiative.org/ (accessed 25.02.25).

140 Dec. 15.42.

141 Decs 16.43–16.46.

142 Resolution Conf. 11.17 (Rev. CoP17).

143 UN General Assembly Sixty-ninth session (2015) Resolution 69/314.

144 United Nations Office on Drugs and Crime (2016) *World Wildlife Crime Report: Trafficking in Protected Species.* United Nations, New York.

145 Doc. 8.6.

146 Doc. 11.20.1.

147 CoP14 Doc. 25.

148 CoP12 Doc. 27.

149 Notification to the Parties No. 2015/039 of 25 June 2015.

150 Resolution Conf. 11.3.
151 Doc. 11.20.1.
152 CoP15 Doc. 24.
153 SC61 Doc. 30.
154 Security Council Resolutions 2136 (2014) and 2134 (2014).
155 UN General Assembly Sixty-ninth session (2015) Resolution 69/314.
156 United Nations General Assembly Seventieth session (2015) Resolution A/RES/70/1.
157 Resolution 2001/12 of 24 July 2001 "Illicit trafficking in protected species of wild flora and fauna".
158 G20 (2017) Annex to G20 Leaders Declaration: G20 High Level Principles on Combatting Corruption Related to Illegal Trade in Wildlife and Wildlife Products.
159 Resolutions 1/3: Illegal Trade in Wildlife and 2/4 Illegal Trade in Wildlife and Wildlife Products.
160 Wright, E.M., Bhammar, H.M., Gonzalez Velosa, A.M. and Sobrevila, C. (2016) *Analysis of International Funding to Tackle Illegal Wildlife Trade*. World Bank Group. Washington, DC. http://documents.worldbank.org/curated/en/695451479221164739.
161 CITES Secretariat (2015) *Corruption as an Enabler of Wildlife and Forest Crime—Joint Statement of the Executive Director of UNODC and the Secretary-General of CITES.* https://cites.org/eng/joint_statement_unodc_cites_on_corruption_wildlife_03112015 (accessed 25.02.25).
162 SC66 Doc. 32.1.
163 Dec. 17.83.
164 Resolution Conf. 17.6.
165 Anon (2019) *Scaling Back Corruption: A Guide on Addressing Corruption for Wildlife Management Authorities*. United Nations, Vienna, Austria.
166 Dec. 19.77.
167 SC61 Doc. 23.
168 Dec. 12.39.
169 Dec. 12.37.
170 CoP13 Com. I Rep. 12 (Rev. 1).
171 Dec. 13.26.
172 Dec. 13.26.
173 CoP14 Doc. 53.1.
174 CoP14 Inf. 56.
175 Notification to the Parties No. 2007/029 of 17 September 2007.
176 Notification to the Parties No. 2008/013 of 26 February 2008.
177 Dec. 13.26 (Rev. CoP15).
178 CoP15 Doc. 44.1 (Rev. 1).
179 SC62 Summary Record.
180 SC63 Summary Record.
181 SC65 Summary Record.
182 Notifications to the Parties Nos. 2015/014, 2015/013 and 2015/012, all of 19 March 2015.
183 Notification to the Parties No. 2016/010 of 11 February 2016.
184 CoP14 Doc. 53.1.
185 Resolution Conf. 10.10 (Rev CoP17) Annex 3.
186 Resolution Conf. 10.10 (Rev. CoP19) Annex 3.
187 Resolution Conf. 14.3 (Rev. CoP18).
188 Lyons, J.A., Daniel, J., Natusch, D. and Jenkins, R.W.G. (2017) *A Guide to the Application of CITES Source Codes*. CITES Secretariat, Geneva, Switzerland.
189 https://cites.org/eng/prog/captive-breeding (accessed 12.02.25).
190 Resolution Conf. 1.6.

191 Resolution Conf. 9.19.

192 Anon (1993) *Report of the United Nations Conference on Environment and Development, Rio de Janeiro, 3–14 June 1992. Volume I Resolutions Adopted by the Conference.* United Nations. New York. A/CONF.151/26/Rev.l (Vol. l).

193 CITES Secretariat (2022) *World Wildlife Trade Report.* CITES Secretariat. Geneva, Switzerland.

194 Lavorgna, A., Middleton, S.E., Whitehead, D., Cowell, C. and Payne, M. (2020) *FloraGuard: Tackling the Illegal Trade in Endangered Plants.* Royal Botanic Gardens, Kew.

195 SC70 Doc. 31.1.

196 CITES Secretariat (2022) World Wildlife Trade Report. CITES Secretariat. Geneva, Switzerland.

197 Harfoot, M., Glaser, S., Tittensor, D., Britten, G., McLardy, C., Malsch, K. and Burgess, N. (2018) Unveiling the Patterns and Trends in 40 Years of Global Trade in CITES-Listed Wildlife. *Biological Conservation*, 223: 47–57.

198 SC61 Doc. 27.

199 AC27 Doc 17 (Rev. 1) Annex 1, AC28 Doc.13.1 Annex, SC66 Doc. 41.1 and SC70 Doc. 31.1 Annex 7.

200 Lyons, J.A., Jenkins, R.W.G. and Natusch, D.J.D. (2017) *Guidance for Inspection of Captive Breeding and Ranching Facilities.* CITES Secretariat, Geneva, Switzerland.

201 Lyons, J.A., Daniel, J., Natusch, D. and Jenkins, R.W.G. (2017) *A Guide to the Application of CITES Source Codes.* CITES Secretariat, Geneva, Switzerland.

202 CITES Secretariat (2022) *New Multilingual Tool to Assist with Inspection of Captive-Breeding and Ranching Facilities and Application of Source Codes on CITES Permits.* https://cites.org/eng/news/new_multilingual_tool_inspection_captive_breeding_ranching_facilities_application_source_codes_cites_permits_31012022 (accessed 25.02.25).

203 Dec. 16.66.

204 AC27 Summary Record.

205 AC28 Summary Record.

206 AC28 Com. 5 (Rev. by Sec.).

207 SC66 Doc. 41.2.

208 Resolution Conf. 17.7.

209 SC77 Summary Record.

210 AC23 Summary Record.

211 Decs 18.102 18.108 and 18.109.

212 SC74 Doc. 28.1.

213 SC77 Doc. 33.8.

214 SC74 Doc. 28.1.

215 Doc. 1.32.

216 Doc. 3.20.

217 Doc. 3.20.

218 Resolution Conf. 3.4.

219 Doc. 7.7.1 Annex.

220 CITES Secretariat (1999) *Memorandum of Understanding (MOU) Concluded between TRAFFIC International, on Behalf of the TRAFFIC Network, 219c, Huntingdon Road, Cambridge, CB3 0DL, United Kingdom and the United Nations Environment Programme, Secretariat of the Convention on International Trade in Endangered Species of Wild Fauna and Flora (the CITES Secretariat), 15, Chemin des Anémones, 1219 Châtelaine.* Geneva, Switzerland. https://cites.org/sites/default/files/common/disc/sec/CITES-TRAFFIC.pdf (accessed 25.02.25).

221 SC38 Summary Record.

222 Anon (2025) *The Green Customs Initiative. The Green Customs Initiative Secretariat, Law Division, United Nations Environment Programme.* www.greencustoms.org/ (accessed 25.02.25).

223 ICCWC (2025) *Tools and Services.* International Consortium on Combating Wildlife Crime https://iccwc-wildlifecrime.org/tools-and-services (accessed 25.02.25).

224 CITES Secretariat (2025) *The CITES Virtual College.* https://cites.org/eng/virtual-college (accessed 25.02.25).

225 CoP16 Doc. 22 (Rev. 1).

226 SC70 Doc. 22.1 Annex 5.

227 CoP18 Doc. 21.3.

228 Resolution Conf. 19.2.

229 SC78 Doc. 21.

230 Doc. 3.21.

231 Resolution Conf. 3.10.

232 Resolution Conf. 5.6.

233 Doc. 6.18.

234 Doc. 9.2.

235 Doc. 10.26.

236 Doc. 9.21.

237 Doc. 10.82 (Rev.).

238 Doc. 11.57.

239 Dec. 11.1.

240 CoP13 Doc. 45.

241 CoP13 Com. II Rep.10.

242 Dec. 14.56.

243 CITES Electronic Permitting Toolkit. Version 1.0. 2010. https://cites.org/sites/default/files/common/cop/15/doc/E15-30-01T.pdf (accessed 25.02.25).

244 Automation of CITES Permit Procedures and Electronic Information Exchange for Improved Control of International Trade in Endangered Species (eCITES). https://cites.org/sites/default/files/eng/prog/e/eCITES_policy_brief.pdf (accessed 25.02.25).

245 Guidelines and Specifications for Electronic Permit Information eXchange (EPIX) of CITES Permits and Certificates (Version as of March 2021). https://cites.org/sites/default/files/eng/prog/e/CITES-EPIX-Guidelines-2022.pdf (accessed 25.02.25).

246 CITES Secretariat (2023) *eCITES: Transforming Global Wildlife Trade Management.* https://cites.org/eng/news/sg-statements/eCITES-transforming-global-wildlife-trade-management (accessed 25.02.25).

247 SC73 Inf. 1.

248 SC78 Doc. 10.

6 CITES
Past, Present, and Future

Books about CITES during its first 25 years were sparse, but since then a number have been published, details of which are provided in Annex 1 of this book. There have also been a number of critiques of the Convention in peer-reviewed and other journals, including some which merit consideration by Parties.[1] This chapter is not intended to review these opinions and commentaries, it is a more personal reflection on the development of the Convention and the successes and failures in its first 50 years. It touches on some of the milestones in the history of the Convention, which are listed in Annex 2 of the book.

Any assessment of the success or otherwise of a Convention such as CITES needs to start with identifying the basic reasons justifying the need for it and how these were articulated in the text of the Convention. Assessing the strengths and weaknesses of the processes and procedures agreed for achieving the Convention's goals and the international cooperation needed to reach them can then be reviewed.

The Convention was developed in the 1960s and 1970s, which were decades of change in public and political attitudes towards wildlife throughout the world. The possibility that some species may become extinct, including because of international trade, caught public attention.

The preamble of CITES make two particular observations. Firstly, that wild plants and animals are valuable in many different ways, they need to be cared for, and peoples and States are and should be the best protectors of their own wild fauna and flora. Secondly, 'certain species' need to be 'protected . . . from over-exploitation' and that international co-operation is essential to do this. The Convention text is vaguer on the identity of the species being over-exploited for international trade and about what constitutes 'over-exploitation'.

6.1 CITES SPECIES

The Convention requires Parties to limit the international trade in species to avoid their endangerment, rather than to maximize the conservation outcome of their actions. Thus, specimens of endangered species listed in Appendix I cannot be allowed in commercial international trade even if this would be advantageous to their conservation. Such a perverse approach has rankled with some Parties and observers but is a consequence of the fact that at the time of its conception, the Convention saw international trade in wildlife largely in a negative light. No doubt if the text was negotiated today, it would have a more nuanced approach.

At the time of the entry into force of the Convention a large number of species threatened with extinction, or likely to become so, had been selected for inclusion in Appendices I and II, based on some basic guidance in the text of the Convention.

DOI: 10.1201/9781003542278-6

The lists of species and the way that they were chosen was considered unsatisfactory, however, and by 1994, Parties had settled on a set of biological science-based criteria for deciding the 'certain species' to be listed in its Appendices.[2] For want of time and resources, no systematic review of the species initially listed in the Appendices has taken place to see if they qualify for listing, despite recognition by the Parties over many years that some do not.

The deterioration of the natural environment has undoubtedly resulted in many additional species qualifying for inclusion in the Appendices since the entry into force of the Convention. It is notable, though, that the additions to the Appendices that have been made over the years are dominated by animal species, which have always commanded greater attention in CITES because of their popularity and media-friendly nature. Some studies have concluded that over 900 threatened species which might merit inclusion in the Appendices are not presently listed by CITES.[3]

Strangely though, most of the species listed in the Appendices are not themselves threatened with extinction or may become so. They are listed in Appendix II under an ancillary provision because they must be subject to regulation in order that trade in the other species included in Appendix II which actually may become threatened can be brought under effective control. Generally, this is because specimens of the former species resemble specimens of the latter to the extent that the enforcement officers who encounter specimens are unlikely to be able to distinguish between them (look-alike species). Whilst this may be of assistance to enforcement officers, the application of full CITES Appendix II controls to tens of thousands of non-threatened species (the precise figure has never been delineated) brings a considerable workload for CITES authorities. At one stage, the USA proposed a two-tier treatment for some aspects of CITES implementation for these two groups of species,[4] but the idea was not pursued and the full weight of CITES provisions is applied to all listed species whether or not they are threatened with extinction or likely to become so.

What continues to develop is the application of the CITES listing criteria to categories of species which had earlier been seen as 'off-limits' for CITES. When the Convention text was finalized, it was not seen as being involved in regulating trade in major resource industries such as fisheries and forestry, where biological extinction is rarely the problem. However, from around 1992, the criteria began to be applied in earnest to tree species used in timber and ancillary industries, and from around 1997 the same applied to fish species. The principal reason for this change in focus was the manifest failure of other bodies charged with managing such species, to halt serious declines in the status of many of the species that they manage, due to international trade.

CITES was not well equipped to address the sustainable management of such species. However, working with partners such as FAO and ITTO, which have good expertise but few means to encourage countries to improve species management, the Convention is finding its place. The international trade in some other groups of species has also yet to be fully explored. Parties agreed that fungi could be covered by the Convention at CoP12 in 2002,[5] but no species have been included in the Appendices since, and there has been little research about whether any species might qualify for inclusion.[6,7] Although there is significant trade in many plant species used for medicinal and aromatic purposes, the impact of international trade on the species used is rather poorly known, and CITES Appendix II has not been applied in many cases.[8]

The listing criteria used by CITES are based on those used in the widely referenced IUCN Red List of Threatened Species, and as such, the science behind them should be sound. Whilst there has often been heated debates about individual listing proposals and which species should be in which Appendix, it seems the settled view of Parties that the criteria for listing species in the Appendices are about right. The reality though is that listing decisions by Parties all too often ignore the listing criteria and are regularly clouded by emotion or driven by political or economic interests.

The inclusion of timber and fish species in the Appendices was resisted for many years by Parties who had a strong economic interest in using these species. On the other hand, lobby group pressure led to the inclusion in the Appendices of a number of popular and photogenic species whose status in the wild did not appear to warrant it. One recent example was the inclusion of the Giraffe *Giraffa camelopardalis* in Appendix II at CoP18 in 2019 despite little evidence to suggest that international trade posed any threat to its survival, and that range States where international trade was allowed reported that the species' abundance was increasing, not decreasing. Should the tendency to ignore the adopted listing criteria continue to escalate, the Convention risks losing credibility and effectiveness.

An issue that has increased in importance over time, as CITES has broadened its focus from preventing trade causing extinction to include sustainable management issues, is that no serious consideration was originally given to including socio-economic considerations in the listing criteria. Codifying them into the criteria would clearly be challenging, but ignoring them is increasingly anachronistic.

A practical complication of the criteria that has proved increasingly problematic is that it tends to be applied to species as a single unit, despite their status and management needs, varying throughout their range. Virtually all species are reasonably rare at the margins of their range relative to core areas of occupation. Such disparities can cause conflict between Parties in different parts of a species' range, as has occurred notably in what is arguably CITES' knottiest problem—the African elephant. In the Southern African States, the species is most abundant, sometimes at carrying capacity, and their management strategy often allowed legal international trade in elephant products (mostly ivory). In contrast, Western and Central African States have seen illegal killing for international trade greatly reduce their own elephant populations, and see demand and international trade in elephant specimens, legal or illegal, as the main reason for their elephant population declines—so they oppose it. Both perspectives are understandable.

CITES has been unable to find a formula to fully bridge this gap. The split-listing of the species that has occurred (with some populations in Appendix I and others in Appendix II) has been difficult and time-consuming to negotiate, complicated to administer, and often left both sides disappointed. Observer NGO groups have expended a huge effort to prevent any international trade in African elephant products. This has brought publicity to the need to find durable solutions to conserve African elephants but has meant that frank dialogue between States involved has been rendered extremely difficult.

Ironically, notwithstanding these challenges and regardless of their significance for CITES as a whole, the decisions about listing species in the CITES Appendices are always a highlight at meetings of the CoP. When they affect charismatic mega-fauna species, they bring drama to the proceedings and attract attention to the work of the Convention from the world's media and the general public. They have been criticized at times for being more showbiz than science, but the Convention would be much duller without them.

6.2 SUSTAINABLE USE

For those species listed in the Appendices, the extent to which extinctions or population declines due to over-exploitation for international trade have been prevented by the Convention is largely undocumented.

Whilst some Appendix I species have become extinct on CITES' watch, most of these extinctions were due to factors other than international trade. A notable exception is Spix's macaw (*Cyanopsitta spixii*). This species became extinct in its native Brazil around 2000, with one of the principal causes being trapping for (illegal) international trade.[9] Around 200 specimens do however survive in captivity, and their movement from one captive breeding facility to another does involve the issuance of CITES permits—a matter that itself has proved contentious.[10] The population status of some Appendix I species that have been the subject of particular attention by the Parties show some encouraging signs. For instance, during the period 1995 to 2015 the best continental population estimates for the African elephant indicate that numbers remained relatively unchanged during that period (Figure 4.3). For rhinoceroses, the IUCN/SSC African and Asian Rhino Specialist Groups and TRAFFIC reports prepared for each meeting of the CoP show that the populations of all rhinoceros species in the wild, except for Sumatran rhinoceros (*Dicerorhinus sumatrensis*), increased between 2007 and 2019. Notwithstanding recent declines in the populations of some species due to, among other factors, the drought in southern Africa, the total number of rhinoceroses in the wild increased by over 28% during this period.[11]

For Appendix II species, the primary obligation of the text of the Convention for Parties is to prevent the species' status declining due to excessive use for trade to the extent that it reaches the critical levels of the Appendix I, and to ensure its ongoing use does not compromise its ability to maintain its role in the ecosystems in which it occurs, throughout its range. Irrespective of the effects of international trade, the range of many species throughout the world has been reduced due to ever-increasing habitat loss in the face of the rising human population and the consequent land use changes to provide for it. Defining the role of a species in its ecosystems is a considerable academic challenge with little prospect of 184 countries quickly agreeing on what this means for the 39,246 species included in Appendix II. Within the CITES machinery, the task of identifying overuse of Appendix II species for international trade has been delegated to the expert opinion of the Animals and Plants Committees. Working in cooperation with the UNEP-WCMC, the committees have developed a well-honed process for flagging up possible overuse.[12] The methodology does not involve consideration of the role that individual species play in their ecosystem, but it is implicit that if the abundance of the species declines, then its role will be impacted as well. The determination of which species from the candidate list, prepared by UNEP-WCMC, and in which Parties, are being subject to detrimental international trade contrary to the treaty then requires the judgement by the science committees, after input from Party and NGO observers at their meetings.

The committees advancing these issues frequently have to tread a fine line. They are under pressure from Parties who seek to defend their own actions in allowing trade and from observer NGOs, some of the leading figures of which are generally opposed to international trade in wildlife under any circumstances.[13] Judging by the reaction of Parties, the Animals and Plants Committees have found a good balance. A more objective analysis of their decisions is, however missing, handicapped by a lack of knowledge about the distribution and status of species in the wild and tools to quantify and monitor it over time. Some attempts have been made to compare trends in the conservation status of CITES-listed species with non-CITES species using the IUCN Red List of Threatened Species. They showed that on average the status of CITES-listed species in the wild continues to decline, but comparing this level of decline with that experienced by non-CITES listed species was confounded by other variables.[14] CITES can and does alter the pattern of international trade in Appendix II species, changing source and market countries. It has avoided many species being transferred to Appendix I because of ongoing documented declines under Appendix II by reducing the volume of trade for many Appendix II species. It remains rather unclear though whether the Convention, by itself, has had a positive effect on the status of species listed in its Appendices.

6.3   REVIEWS OF CITES AND STRATEGIC PLANNING

In 1993, Zimbabwe, on behalf of a number of Southern African States prepared a trenchant critique of the Convention which it sent to UNEP and the Secretariat. It bemoaned many aspects of the treaty and called for radical revision. The document does not seem to have been deployed in CITES forums (a revised version was placed in the public domain in 2016),[15] but it may not have been a coincidence that it arose just before the first, and so far, only, structural review of the workings of the Convention.

At a meeting of the Standing Committee in 1994, Canada suggested that a review of the general evolution and the implementation of the Convention should be conducted by an independent body.[16] This proposal was approved by CoP9 in 1994.[17] The review's stated aim was to assess the effectiveness and efficiency of the provisions of CITES and the way it was implemented to achieve the objective of the Convention. It was conducted in 1996 by a consultant hired by the Secretariat, Environmental Resources Management, and was titled 'How to improve the effectiveness of the Convention'. After circulation to the Parties for comment[18] and its clearance by the Standing Committee,[19] it was submitted to CoP10 in 1997.[20] The reaction from Parties was mixed,[21] and it seemed clear that the preparation of the report was hampered by insufficient time and funding, given the complexity of the task. Nonetheless, its recommendations in the form of a 22-point action plan[22] were adopted by CoP10[23] together with around 30 other detailed decisions arising from its findings. Amongst the more significant of the latter were instructions for the Standing Committee to lead the development of a medium/long-term framework document for the work of the permanent Committees and to give consideration to the development of performance indicators for the Convention.[24]

This led to the preparation and adoption of a first Strategic Vision for the Convention covering the period 2000–2005,[25] which initially, at least, brought much needed structure and direction to the work of the Convention's bodies. This Strategic Vision was extended to 2007 and has been followed by subsequent revisions: 2008–2020[26] and 2021–2030,[27] which have been endorsed with declining degrees of interest and enthusiasm by meetings of the CoP. For the second and third iterations of the Strategic Vision, Parties developed indicators for the objectives set. Those for 2008–2020 were so extensive and detailed that they proved unusable.[28] Indicators for 2021–2030 were added at CoP19 in 2022, three years after the Strategic Vision itself was adopted.[29] They were more focussed than previous attempts and linked to the information requested in the format of the Parties' implementation reports. However, an indicator to measure progress against arguably one of the most critical objectives of the Strategic Vision—that the Appendices correctly reflect the conservation status and needs of species—could not be determined, and the Standing Committee agreed not to propose one,[30] reflecting a long-standing difficulty to determine if the Convention is addressing the correct species.

The Strategic Vison 2008–2020 included a vision statement, which sought to put the work of the Convention into the context of broader concerns about species conservation globally, by linking it to wider objectives of the overarching Aichi Biodiversity Targets of the CBD, and that for 2021–2030 made this linkage more succinctly:

CITES VISION STATEMENT

By 2030, all international trade in wild fauna and flora is legal and sustainable, consistent with the long-term conservation of species, and thereby contributing to halting biodiversity loss, to ensuring its sustainable use, and to achieving the 2030 Agenda for Sustainable Development.

The 1996 external review of the effectiveness of the Convention had its limitations, but it did lead to much more coherent thinking about the role of the Convention and the way that it is implemented.

At CoP18 in 2019, the Democratic Republic of the Congo, Namibia, South Africa, and Zimbabwe proposed a second comprehensive review of the Convention with the aim this time of improving the equitability of the Convention, especially in regard to the rights of rural communities and indigenous people over their own natural resources including wild animals and plants.[31] The tone of the proposal was similar to the 1993 critique of the Convention by Zimbabwe and echoed an ongoing debate between African Parties broadly in favour of the use of wild species and African Parties, supported by many observer NGOs, broadly opposed. This difference of approach was subsequently reflected in the debate about the proposal at the CoP18.[32] Although the CoP agreed on an instruction to the Standing Committee to reflect on the need for a further review of the implementation of the Convention, the issue was quietly buried.[33]

6.4   GLOBAL SUPPORT FOR THE CONVENTION'S OBJECTIVES

Helped by increased recognition of the importance of the world's natural environ-mental generally, after its entry into effect, the Convention quickly gained adherents and reached 100 Parties by 1989. Progress in recruiting new Parties has been slower since around 2008, but nonetheless the number of Parties reached 185 by 2025.

Just nine United Nations member States from different regions are not yet Party to the Convention: the Democratic People's Republic of Korea and Timor-Leste in Asia; South Sudan in Africa; Haiti in Central and South America and the Caribbean; and the Federated States of Micronesia, Kiribati, the Marshall Islands, Nauru, and Tuvalu in Oceania. None of these States are major international traders in wildlife. A number have made overtures about joining the Convention. Additionally, there are a small num-ber of non-United Nations members and UN observers who are not a Party (the Holy See, the Cook Islands, Niue, and Palestine) and a handful of dependent territories of UN member States who have decided not to apply the Convention—even if the State upon which they depend is a member in each case. Universal membership of CITES is within reach, but this will depend on proactive action by the CITES Secretariat, which is cur-rently fully occupied with more pressing priorities. So far, the only Party to denounce the Convention and leave was the United Arab Emirates in 1988. They rejoined two years later. Some southern Africa States have threatened to leave on a number of occa-sions, but have not done so. Although multilateralism generally has been questioned in recent years, so far CITES has been largely unaffected by this uncertainty.

Amongst the Parties, examination of the actions of CITES over the past 50 years reveals that the USA has been the instigator or a major player in most of the key policy developments. This reflects a consistent policy of engagement by the USA and a series of competent and dedicated lead officials. India hosted CoP3 in 1981 and was a major force in the early years, but its role has declined considerably since. Also, Mexico and Canada were and remain key Parties in the development of CITES' global policies. The EU and its member States have been very active participants in CITES and the principal donor of voluntary financial contributions to the Convention. The ceding of competence of the EU's member States to the REIO and the abolition of the internal borders within the EU was a political reality which needed to be accommodated by CITES. Whilst some concerns about its implementation of the Convention persist, the EU's member-ship of CITES can be seen to have had a positive impact on the Convention.

The dominant voices in CITES have certainly been from Europe and North America. As previously mentioned, Africa's position in CITES has been character-ized by differences of approach to wildlife trade between Southern African States and Western and Central African countries, led by Kenya. In Southern Africa, Graham Child, Rowan Martin, and Willie Nduku of Zimbabwe stood out as key advocates of the views of their constituency, and in Kenya this role was played by Richard Leakey and latterly Patrick Omondi. Japan has been a somewhat reluc-tant supporter of CITES, but the fact that it became a Party to the Convention and reflected a more utilitarian approach to wildlife use led to other Parties having to face up to some challenging questions. Often out-voted on policy issues, Japan has sought to limit the impact of CITES by adopting a consistently aggressive approach to cost-cutting in budget negotiations. China's role in CITES has mirrored its rapid economic development over recent years. Initially playing a rather marginal role, it has become one of the major consumers of wildlife and with its large human population and huge international trade, a significant player in CITES. Although regularly criticized by many NGO observers, it has made significant strides in implementing the Convention and has latterly been an active participant in its policy development. Whilst they have had their CITES champions over the years, such as Victoria Lichtschein of Argentina, Rod Hay of New Zealand, and Hank Jenkins of Australia, the regions of Central and South America and the Caribbean and Oceania have generally played a lesser role in the development of the Convention globally.

6.5   INFLUENCE OF OBSERVERS ON CITES

The provisions of the treaty provide for the participation by any body or agency technically qualified in protection, conservation, or management of wild fauna and flora (including NGOs) at meetings of the CITES CoPs, which was a pioneering move at the time of its entry into force. CITES subsidiary bodies gradually applied similar participation rules to those which applied to the CoPs. Since their creation of the Animals and Plants Committees in 1986, the chairs regularly invited NGO observers to attend meetings of their committees in view of the expertise that they could bring. This was considered rather less sensitive given the largely advisory, rather than decision-making roles that the committees have. In 2018 the science committees' Rules of Procedure were changed to dispense with the need for an invitation from the Chair, even if this had become a formality and participation was largely open. In contrast, it was not until 2003 that observer NGOs were able to apply for an invitation to attend meetings of the Standing Committee.

Initially, the number of conservation NGOs observers at CITES meetings was rather limited, but it picked up from around CoP5 in 1985 and has since grown rapidly (Figure 6.1).

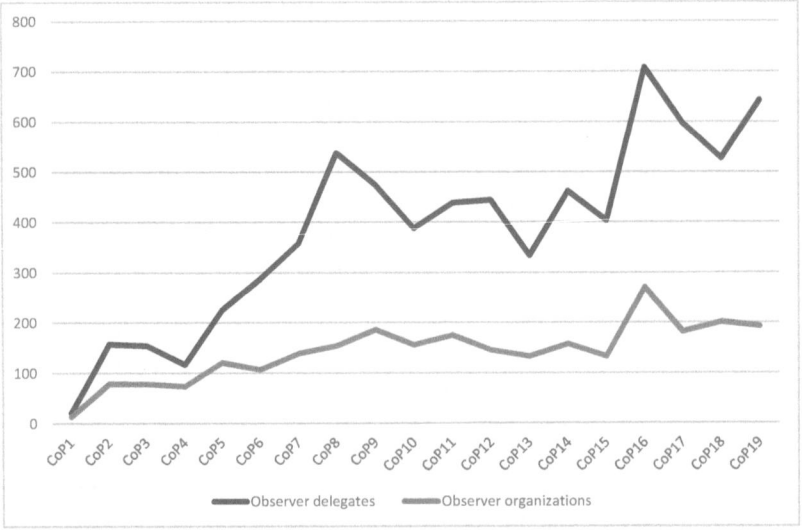

FIGURE 6.1   Number of observer organizations and observer delegates present at meetings of the Conference of the Parties (Courtesy of Dan Challender).

The rise of World Wide Web in the 1990s and of social media in the late 1990s and early 2000s provided enhanced campaigning and fundraising possibilities leading to the development of a new model of conservation NGO which depended, not on membership and expertise, but on slick publicity and marketing. A number of these observer NGOs are focussed on animal welfare concerns, rather than species conservation. At CoP5 in 1985, Israel called on the Parties to endorse, in principle, the concept of a Convention for the Protection of Animals which would establish international standards on the procurement, transport, and captive maintenance of animals.[34] The CoP voted by 42 votes to 5 that the proposal was ultra vires, but in the absence of any such a treaty being developed, Israel and some other Parties together with some observer NGOs, have since sought to apply some of the thinking behind this idea in the forum of CITES.

Whilst observers from wildlife business interests are able to participate in CITES meetings, their influence has been much more muted; exceptions are the crocodilian leather, ornamental fish and caviar industries, and latterly, those involved in the musical instrument use and trade. Observers from civil society in developing countries do not generally have the means to attend CITES meetings, and in contrast to other biodiversity MEAs such as the Convention on Biological Diversity (CBD), participation at CITES meetings by indigenous peoples and local communities is very limited. Some Parties, such as Canada and Denmark, do include representatives from indigenous peoples and local communities in their delegations, and they have spoken decisively on issues such as the potential transfer of polar bears, *Ursus maritimus*, from Appendix II to Appendix I.[35]

Coming mostly from Europe and North America, observer NGOs participating at CITES meetings often hold similar views and perspectives. The influence of Western conservation and animal welfare observer NGOs on the functioning of the Convention's governing and advisory bodies has grown as resources available to governments have declined. They have been labelled 'conservation influencers' by some critics.[36] What is clear is that these observer NGOs have staffing and budgets far in excess of the CITES Secretariat and the government departments that they seek to lobby. Their conduct and influence has drawn adverse comment from Parties and commentators,[37,38] yet evidence suggests that they are being successful in impacting decision-making and adjusting the priorities of the Convention.[39]

Chairs at sessions of the CoP and in the CITES committees habitually gave observer NGOs the opportunity to speak on all issues, but lengthening meeting agendas made this increasingly difficult. A number of mostly European and North American NGO observers created the Species Survival Network coalition in 1992, enabling interventions made at meetings of the CoP on behalf of a large group of like-minded observers, and giving them a better chance for their voice to be heard in debates. Latterly however, these NGO observers have instead briefed delegates directly and given them prepared statements to read out at CITES meetings. Lack of resources in government departments responsible for CITES has enabled observer NGOs to take the initiative to ghost-write documents for Parties to submit to meetings of the CoP and CITES subsidiary bodies. This often enables hard-pressed government officials to create popular headlines at home, but results in a lack of real engagement and buy-in for follow-up. It may have contributed to overcrowded meeting agendas and to new policies which are often rather shallow and result in a lack of meaningful change. As NGO observers increased their number and influence threaten to create an imbalance.

The text of the Convention has provisions to exclude certain observers from a meeting of the CoP if at least one-third of the Parties present object to their presence. Permission for observers to participate at meetings of the Standing, Animals and Plants Committees may be withdrawn at any time if the members present agree. In practice these possibilities are rarely, if ever, used. Parties have taken some steps to moderate the influence of observers at meetings: for example, by limiting the number that can participate in in-session working groups to those with expertise in the matter to be discussed and ensuring that they do not outnumber official delegates from Parties. Calls for a broader code of responsibility for NGOs at CITES meetings[40] were not taken up by Parties. NGO sponsorship of Party delegates to participate at meetings of the CITES CoPs was an increasing concern, and at CoP17 in 2016, Parties agreed to request donors (and recipient delegates) to declare if funding had been provided for attendance at a particular meeting of the CoP. This was in order to avoid the perception of inappropriate influence and to ensure that the Convention is seen to be operating in a fully open and transparent manner.[41] At CoP18 in 2019, 24 delegates were reported as receiving such bilateral sponsorship, falling to 13 delegates at CoP19 in 2022.[42] It is unclear if this is the true extent of donor support paying for delegates to attend meetings of the CITES CoPs or if this support is provided for other CITES committee meetings where similar provisions do not apply.

Whilst there is no question that observer NGOs bring the oxygen of publicity to CITES which generates political support and funding, there is an increasing need to find a better balance between transparency, civil society engagement and respect for the sovereign rights of delegates representing all the citizens of their respective countries.

6.6   CITES MACHINERY

For the first ten years of its existence, CITES subsidiary bodies developed in a rather ad hoc and unstructured way. The creation of the existing structure of intersessional committees in 1987 was a turning point in the global functioning of the Convention. The Standing, Animals, and Plants Committees form a well-balanced triumvirate, which functions efficiently and has served the Convention well. Subsequent attempts to create additional intersessional bodies on compliance, enforcement, and rural communities have been rebuffed. Whilst the intersessional bodies function well, the workload delegated to them (and the Secretariat) by the CoP has grown rapidly since CoP15 in 2010. Much of this extra burden, largely in the form of CoP decisions, has resulted from a lack of prioritization by the CoP. All issues addressed seem to have been given the same importance. Maximum capacity was probably reached in 2019, and now the volume of work greatly exceeds the capacity of the committees and the Secretariat, constrained as they are by financial and human resources. The result has been increasingly rushed and shallow decisions. In 2024 and 2025, despite the best efforts of the Chair, the Standing Committee failed to complete its meeting agendas. The response of the Secretariat, in consultation with the Chair of the Standing Committee, has been to extend the duration of meetings, but reduce the time for debate.[43] It seems unlikely that this will result in better decision-making, and it will certainly handicap those Parties for whom English, French, or Spanish are not their native tongue. Radical action is needed to avoid the Parties over-extending themselves. As in other fields related to the implementation of the Convention, all proposed new activities have their strong supporting Parties, and other Parties hesitate to oppose, for fear of appearing too negative or being wary that their own special interests may also fall victim to rationalization later. An initiative from the Secretariat is needed.

The Convention's governing and advisory bodies rely mostly on face-to-face meetings. These have proved increasingly popular, with attendances rising year-on-year. International travel to meetings and events and to visit Parties is a significant part of the Secretariat's activities. However, little account has been taken of the environmental impact of holding CITES meetings, nor of the actions of the Secretariat on the world's climate. In 2007, the United Nations adopted a climate-neutral programme and at first the Secretariat set aside 10% of its travel budget to offset the carbon footprint resulting from staff travel.[44] UNEP also produced guidance on holding 'green meetings',[45] but lack of capacity in the Secretariat has meant that this has rarely been referred to in CITES. As explained later in this chapter, when Parties ordered a review of climate change, its focus was on the impact it may have had on CITES' objectives, but the impact of the Convention's activities on the climate has rarely been considered. As an environmental organization, CITES could and should be doing better on this issue.

6.7   CITES SECRETARIAT

The secretariat of CITES has a broad mandate for such an international convention and for the most part has used it wisely and played a significant role in the development of the Convention. Whilst most bodies of its kind plead poverty, it is fair to say that the CITES Secretariat has been under-resourced, in relation to the tasks that the Parties ask it to conduct, throughout its existence.

Although the budget for the Secretariat has increased over the years, it has not taken into account the huge increase in the number of Parties to be serviced, nor the increasing workload for the Secretariat decided by the CoP. The core staff size of the Secretariat has remained largely the same for many years. The recruitment of unpaid (unless they are in receipt of stipend from a sponsor) interns and temporary staff for specific tasks, using external donor funds, has increased, and this has relieved some pressure on core staff. There have also been occasional loans of staff from Parties— although these have not always been a success.

The Secretariat has prepared a broad costed programme of work for the CoP since CoP14 in 2007, but it is in the adopted decisions and resolutions that the full scope of CITES' (and the Secretariat's) activities for each triennium are detailed. Proponents of decisions and resolutions submitted to meetings of the CoP are requested to prepare an estimated budget for their proposals, but many do not, and those that do have limited skills to cost out the activities that they propose. In any event, these esti-mates are not taken into account of when the budget for the triennium is determined. Negotiations over the budget take place in closed sessions of a CoP budget working group which meets concurrently with the main business of the CoP meeting. This disconnect means that Parties adopt policies which take little account of available resources in staff or finances.

Virtually all the actual activities undertaken by the Secretariat rely on external funding from donors. Although it has never had a professional fundraiser amongst its core staff, the Secretariat has been remarkably successful at raising funds, aided no doubt by the public popularity of the subject matter. In recent years estimates of donor funding required to complete the tasks assigned to the Secretariat varied from USD 7–20 million per year, and during the period 2016–2018 for example, around 80% of this was actually raised.[46] Fundraising by the Secretariat has relied mostly on donor support from governments in Western States. Although the CoP classi-fies planned activities in its adopted costed programme of work as of core, high, medium, or low priority, donor funds are very rarely given unallocated. They are normally earmarked for specific subject activities of interest to the donor, regardless of the priority of the subject accorded by the CoP. This circumstance does not result in the best outcome for CITES as a whole and reinforces a perception in some quar-ters that the Convention is driven by the small group of Parties who make donations.

The Secretariat has made up for some of these weaknesses by having a par-ticularly loyal and dedicated staff complement. Many staff have remained in post for numerous years, and whilst this may handicap innovation, for a body which is charged mostly with implementing decisions made by others, this is arguably of less importance. It also retains institutional memory, which is particularly relevant given the wide range of calls made on the Secretariat.

The first duty of the Secretariat is to arrange for and service meetings of the Parties and at this the CITES Secretariat has excelled. As a consequence, it has regularly been requested to loan conference organizing staff to other biodiversity MEAs for the duration of their meetings to spread good practice. This effort was led for 24 years by Jonathan Barzdo, who previously headed two of the Secretariat's key implementation partners: the Wildlife Trade Monitoring Unit and TRAFFIC.

For most of its existence the Secretariat possessed no specialist press or com-munications officer, and staff with other specialisms have done this in their spare time. Nevertheless, the Secretariat has maintained a high public profile for CITES, belying it size.

Relations between CITES Parties and UNEP over the budget for, and manage-ment of, the Secretariat have been turbulent since the early days of the Convention. UNEP's attitude to its responsibilities has been inconsistent, often depending on the priorities of its executive director at the time.[47] In recent years, the burden on the Secretariat of operating under UN regulations and rules has grown, particularly since the arrival of the UN resource planning system 'Umoja' in 2015. The UN regu-lations and rules that apply may well be appropriate for large and complex organiza-tions but are over-burdensome for a small team like the CITES Secretariat, resulting in inefficiency and poorer service for the Parties. There may be merit in exploring more flexible arrangements as some other UN organizations have done.

The Secretariat has met many challenges since its establishment, such as recurrent shortage of funds and resources, personnel issues and institutional relationships, but the current overwhelming volume of work is intractable. The situation is getting worse, not better, and Parties need to rationalize and better prioritize the work that they instruct their Secretariat to do.

6.8 A SINGULAR FOCUS

A distinguishing feature of the attitude of the CITES Parties during the last 50 years has been their reluctance to stray outside the mandate of the Convention.

At CoP8 in 1992, on an initiative from Botswana, Malawi, Namibia, and Zimbabwe, Parties recognized that commercial trade in wildlife may be beneficial to the development of local people when carried out at levels that are not detrimental to the survival of the species in question.[48] This landmark resolution subsequently led to wider discussions about the importance of wildlife trade for the livelihoods of people, particularly the poor and those in rural communities, including indigenous and local communities. At CoP13 in 2004 there was a discussion about the possible listing of the plants known as Devil's claw *Harpagophytum* spp. in the Appendices, a species of considerable importance to marginalized human communities. Arising from this debate, Australia proposed that the CoP should recognize that the implementation of CITES-listing decisions should take into account potential impacts on the livelihoods of the poor.[49] Ironically, in the end, it was decided not to include Devil's claw in the Appendices, but Australia's proposal was adopted and set in train a string of workshops, working groups, factsheets, case studies guidelines, and toolkits on the subject of CITES and livelihoods.[50] Yet despite all the effort and warm words, there were very few changes of substance agreed to CITES implementation policy to address this issue, and most of the agreements provided soft guidance to Parties at best.[51]

On a related theme, proposals by Namibia, the United Republic of Tanzania, Zambia, and Zimbabwe to establish a Rural Communities Committee in the CITES machinery to provide guidance and advice to Parties about the potential social impact of amendment proposals, decisions and resolutions of the CoP on indigenous peoples and local communities[52] also failed to gain traction. Most Parties considered that these issues were best addressed through national mechanisms or through the participation of observers at meetings of the Parties. They were against affording rural communities a higher status than other stakeholders and the promotion of socio-economic considerations in a disproportionate way.[53]

At CoP10 in 1997, Argentina, New Zealand, and the USA submitted a discussion document on trade in alien species,[54] and as a consequence three decisions were agreed which were later turned into a resolution on the subject.[55] In the resolution, it was recommended that consideration be given to the problems of invasive alien species when developing national legislation and regulations that deal with the trade in live animals or plants; that exporting Parties should consult importing Parties when considering authorizing the export of potentially invasive species, and that synergy and cooperation might be considered with CBD the IUCN/SSC Invasive Species Specialist Group in relation to alien invasive species. In the course of the latter activity, the Animals and Plants Committee concluded that the contribution of CITES to address threats posed by alien invasive species was likely to be very limited and that the practical utility of further work was questionable, given the means and resources that CBD had already deployed on the issue. Further, they considered that CBD should provide the necessary information and guidance to CITES, and not the other way around.[56] The active engagement of CITES with external partners on the matter of invasive alien species was discontinued.[57]

Other biodiversity MEAs have extensive work programmes on the impact of climate change on their remit. At CoP15 in 2010, on a suggestion from the Secretariat, the CoP asked the Animals and Plants Committees to identify the scientific aspects of the provisions of the Convention and of resolutions of the Conference of the Parties that are actually or likely to be affected by climate change and for the Standing Committee to report to CoP about the issue.[58] When the matter returned to CoP, however, it was agreed that the current provisions of the Convention and of resolutions of the Conference of the Parties were sufficiently comprehensive and flexible to take into account the implications of climate change for science-based decision-making and that no further action was required.[59]

The Secretariat got involved with the issue of wildlife diseases in 2011, when invited to become a 'core affiliate' of the Convention on the Conservation of Migratory Species of Wild Animals (CMS) Scientific Task Force on Wildlife Diseases, which CMS had jointly established with FAO. When this news was passed to the Animals Committee, they recognized the importance of wildlife diseases and the linkages with CITES, but believed that this matter was best addressed by other organizations, such as the World Organization for Animal Health (WOAH).[60] This view was supported by the Standing Committee, who instructed the Secretariat to downgrade its engagement.[61]

In the feverish aftermath of the declaration of COVID-19 as a global pandemic in 2020, some raised concerns about the role of wildlife trade generally, and that regulated by CITES specifically, in the spread of the virus. There were calls by observers for a ban on commercial international trade in live and fresh terrestrial wildlife,[62] despite a lack of evidence that the SARS-CoV-2 virus had been spread by this trade. Other observers proposed amending the CITES treaty to address wildlife trade that poses a threat to human and animal health.[63] When the CITES Parties themselves got to talking about the issue, they adopted a much more measured approach, calling for a review of the matter by the Standing Committee based on a report from the Secretariat.[64] Although this process is still ongoing, it seems unlikely that the Parties will take an active part in combatting the risk of zoonotic disease emergence and spread in future.[65]

At CoP17 in 2016, host country South Africa and the USA submitted a discussion document on youth engagement in CITES.[66] A resulting resolution encouraged Parties to promote this.[67] Singapore took the lead in trying to establish a CITES Global Youth Network with a focus on increasing youth involvement in wildlife trade-related issues, particularly action against illegal wildlife trade.[68] Whilst the initiative has some cheerleaders amongst Parties, it has not so far been woven into the implementation of CITES.

At CoP19 in 2022, host country Panama proposed efforts to promote gender equality and gender mainstreaming in CITES. The proposal proved rather contentious. After a secret ballot with 73 Parties in favour and 29 against[69]—just above the two-thirds majority required, a resolution on gender and international trade in wild fauna and flora was agreed.[70] The development of a CITES Gender Action Plan which was called for at CoP19 in 2022 under decisions accompanying the resolution also caused some concerns about the pertinence of some of its proposed features and its impact on the core business of the Convention.[71]

Parties have consistently shown a conservative approach to efforts aimed at broadening its engagement into issues that were not considered core business. On the one hand this approach has maintained a clear focus on the implementation of CITES provisions, but on the other, it has led to a detachment from wider efforts to conserve and sustainably use biodiversity and a failure to consider wildlife trade in the wider socio-economic milieu in which it exists.[72]

6.9   A CONVENTION WITH TEETH

The CITES compliance procedures, and the compliance measures subsequently applied, have undoubtedly played a decisive part in improving implementation of the Convention, even if much remains to be done. Whilst dialogue and discussion have resolved many of the cases of poor implementation highlighted by the Secretariat under the provisions of Article XIII of the Convention, the list of Parties that have been subject to its ultimate sanction, a recommendation to suspend trade in CITES specimens is still very long (Table 5.1).

When reviewing the list of Parties that have been the subject of recommendations to suspend trade, one striking feature is that almost all are developing countries, which are rich in biodiversity and generally exporters of specimens of CITES species. Despite the emphasis on international co-operation in the preamble of the Convention and the fact that CITES trade transactions require an importing and an exporting Party (and sometimes a transit Party as well), responsibility has often fallen only on the exporting Parties. In the eyes of some, this not only reflects the imbalances of power and the structural inequities of the Convention but also may only result in the demand for illegally sourced specimens merely being displaced from one source country to another.[73]

This perceived imbalance is compounded by the fact that compliance matters are decided in the Standing Committee. Yet until very recently,[74] no financial support was available to permit the attendance at Standing Committee meetings of Parties whose compliance was under review. Exporting Parties—mostly developing countries—are those most likely to be unable to be able to attend Standing Committee meetings for financial reasons. In the past they have often therefore been deprived of the opportunity to put their case in person before a suspension of trade recommendation was agreed. In 2016, in an effort to try and promote positive incentives to comply with the Convention's provisions, the Secretariat proposed that it establish a Compliance Assistance Programme.[75] Although approved by the Parties at CoP18 in 2019,[76] with ambitious objectives[77] it is yet to be fully realized.

The CITES compliance procedures have been widely praised and undoubtedly have been successful in engendering corrective measures by the Parties to whom they have been addressed. However, in too many cases this has not been before the ultimate sanction of a recommendation to suspend trade has been invoked. Most of the targets have also been exporting Parties, largely developing countries. A more balanced and equitable approach to compliance and greater emphasis on preventative action, perhaps through the new Compliance Assistance Programme, would improve the CITES compliance procedures still further.

6.10   WILDLIFE TRADE—LEGAL AND ILLEGAL

Reported legal trade in specimens of CITES-listed species between 2011 and 2020 has been calculated at over 1.3 billion individual organisms (1.26 billion plants and 82 million animals) and an additional 279 million kg of products reported by weight (193 million kg of plants and 86 million kg of animals).[78]

During the period 2015 and 2021 seizures made by enforcement authorities involved roughly 13 million items by number and nearly 17,000 tons by weight. These involved around 4,000 plant and animal species, approximately 3,250 of them listed in the CITES Appendices.[79] It seems likely that these figures are an underestimate as collecting such information is challenging.

The financial value of wildlife trade is not often mentioned in CITES forums, as the focus is on the impact on the species, but in an increasingly monetarized world, perhaps this needs to be reviewed.

Estimating the financial value of wildlife trade, including international trade regulated by CITES, is fraught with difficulty and published estimates typically lack details of methodology and are difficult to evaluate. They can vary hugely depending on whether they try and estimate the value of the whole supply chain or just document it at a certain point, such as the point of initial export from the country of origin. There are strong incentives to over-exaggerate the financial value of the trade, legal or illegal, which is linked to political priorities afforded to the issue. Public organizations concerned about their core funding, officials whose careers are involved and NGOs seeking to raise funds, have obvious incentives to maximize estimates of the value of the trade.

The most credible estimates of the financial value of wildlife trade come from intergovernmental bodies. The value of legal trade in CITES-listed species during the period 2016–2020 was estimated at USD 1.8 billion for animal exports and USD 9.3 billion for plant exports. These figures though only refer to the value at the point of export/import (for animals) or point of sale (for most plant data) and not the full value throughout the supply chain.[80]

The clandestine nature of illegal trade means that it is always likely to be more difficult to estimate its size and financial value. It is clear that illegal trade involving products from the forestry and fishing sectors is far larger than from other types of wildlife. Analyzing previous published studies, UNEP-INTERPOL estimated the annual 'loss of resources' from plants and other wildlife (excluding illegal logging and illegal, unreported, and unregulated fisheries) as 7–23 billion USD in 2016.[81] This includes 'poaching' as well as 'trade' generally. The same report said that forestry crimes including corporate crimes and illegal logging accounted for an estimated USD 51–152 billion per year and illegal fisheries an estimated USD 11–24 billion per year. In 2023, INTERPOL stated that the black market for illegal wildlife products ('poaching and the illegal wildlife trade') was worth up to USD 20 billion per year.[82] In 2019 a World Bank Group report estimated the value of illegal logging, fishing, and wildlife trade at USD 1 trillion or more per year, but 90% of this value was from estimating ecosystem services, not currently priced by the market. There are real difficulties comparing figures on the value of illegal trade. Whatever the precise figures, illegal wildlife trade, including in CITES-listed species, is big business.

Financial rewards from legal wildlife trade can in some cases play a significant role in local and national economies. Illegal trade, by contrast, poses a threat to the conservation of species and deprives local communities and governments of revenue. In some cases, it also fuels corruption in public institutions and poses a security threat at national and regional level.

6.11   THE NEXT 50 YEARS

When the Convention came into force, the world's human population was 4 billion. Today it is just over double that figure and by 2075 is expected to be around 10 billion.[83] The increase in the value of world trade (exports and imports) from 1975 to 2025 comfortably eclipses the growth in human population—increasing from around USD 1,000 billion to USD 25,000 billion.[84] Coupled with this growth in trade has been a rise in disposable income of citizens, particularly in key wildlife consumer countries. According to the World Bank, in China, from 1975 to 2023, GDP per capita (at constant 2015 USD values) rose from 337 USD to over 12,000 USD.[85] The Convention has persisted, and indeed prospered, during a time of enormous change in the demographic and economic background to wildlife trade.

Such is the dizzying speed of current changes in human society, particularly in the field of information and communication technology that it is difficult to predict what may happen in 50 years' time.

It is hard to imagine, however, that international trade in specimens of species taken from the wild will not decline precipitously by 2075. Three factors point in this direction.

Firstly, the distribution of, and abundance of, wild species has already been reduced significantly since 1975. This has been caused by changes in land and sea use, direct exploitation, climate change, pollution, and invasion of alien species. All driven by the increase of the human population and its activities. On present trends, this will get worse in the coming 50 years despite the increasing public awareness. Unless action is taken to reduce the intensity of drivers of biodiversity loss, around 1 million species may be facing extinction.[86] This would affect the supply of specimens of species of the kinds included in the CITES Appendices—largely vertebrates and plants.

Secondly, over the past 50 years there has been a significant increase in the degree of 'privatization' of wildlife resources used in international trade. Species, including CITES species, that were formerly harvested from the wild under differing ownership arrangements are now much more frequently bred in captivity in the case of animals, or artificially propagated or grown on plantations, in the case of plants. Often these production facilities are detached from the wild populations of the species they use. Freed from the vagaries of nature, this type of development provides more stable sources of supply of the specimens of the species concerned, with better quality control, but often provides no incentives for the conservation of the wild populations.

By concentrating the benefits of trade in the hands of a restricted number of people or companies, it creates incentives for them to defend their place in the market from competition from wild-sourced specimens. Production and trade in specimens produced through biotechnology and other technological tools such as three-dimensional printing can also be expected to increase strongly in the future. Parties have been exploring the ramifications of such developments for the Convention since it was first raised by the USA at CoP17 in 2016.[87] No firm conclusions have been drawn yet. The impact on wild populations of trade in CITES-listed species from non-wild sources may be triple-edged. It could reduce prices for wild products, or it could generate incentives for illegal trade by providing an avenue through which wild specimens could be 'laundered',[88] and remove a commercial incentive that could be used to conserve wild species. Furthermore, as the availability of captive bred, artificially propagated, nursery-grown, or synthetic specimens of CITES-listed species grows, so the demand for specimens from wild source may increase as a result of the latter's scarcity and marketing of them as 'authentic' wild specimens. Parties have tried to forge links between ex situ captive breeding and artificial propagation facilities, and conservation the species they use in their natural habitat. A resolution was adopted encouraging such cooperation at CoP13 in 2004,[89] but when a more detailed study of the relationship between ex situ production and in situ conservation was proposed at CoP14 in 2007,[90] the CoP was opposed, with some Parties feeling that the CBD was a more appropriate forum to explore the issue.[91] As the percentage of international trade in specimens of CITES species from non-wild sources increases, there may be scope for Parties to revisit this issue, perhaps in conjunction with CBD.

Thirdly, although it is more difficult to judge its perennity, the current societal trend in many wildlife market countries is moving to oppose the use of animals in trade in general. NGOs and other observers, mostly from Western civil society groups, have gradually got the upper hand in debates about use of species within the CITES forum. If this trend continues, importation of wildlife to many consumer States may be limited or prohibited, regardless of any conservation concerns or impacts.

Hopefully, the scientific knowledge about the distribution, status, biology, sustainable management of species being used for international trade should improve during the next 50 years. A host of new techniques, such as bioacoustics, eDNA and miniaturized GPS transmitters are rapidly improving our knowledge of species and their biology. Certainly, artificial intelligence should allow the available information to be brought together much faster than at present so that decisions, particularly on non-detrimental findings, can be quickly informed by all the known facts.

However the international trade in wildlife develops, national enforcement of laws implementing the Convention's provisions will remain very important. Currently, 90% of all global trade in goods is seaborne, but of the 800 million containers shipped each year, less than 2% undergo inspection.[92] Improved risk-assessment and technological developments should allow more rapid scanning of suspicious shipping containers. DNA-based identification tools are likely to make it much easier to rapidly identify the species from which meat, powder, and similar products are derived. This should facilitate better control and enforcement.

The institutional organs of CITES are well designed, but the way they work is likely to be radically altered. Machine interpretation and translation should allow easier communication between Parties and in many more languages than can be catered for at present. Although Parties are not keen on the idea at present, online meetings of CITES are likely to become the norm, particularly for technical issues.

6.12   FINAL THOUGHTS

When CITES came into effect 50 years ago, some hailed it as a 'Magna Carta for wildlife'.[93] With its limited mandate in terms of the numbers of species to which it applied and the threats that it addressed, it was never likely to live up to such a billing, but its accomplishments would surely have pleased those who conceived it. This may seem counter-intuitive, given that its direct impact on species has not been measured, but its achievements should also be seen in a wider context. Despite its small size, it has contributed significantly to the mainstreaming of public awareness about the importance of nature to human well-being, and the need to prevent demand through international trade posing threats to the survival of species and the functioning of the ecosystems in which they live. It has demonstrated that States can reach consensus on actions to conserve nature and can hold their peers to account when they fail to live up to the standards which have been decided upon. If the rigour of CITES could be applied to all such intergovernmental agreements covering the environment in which we all exist, the world would be a better place. As it is, our human species seems to take nature for granted, despite its importance to our own well-being and the obvious signs that it needs better stewardship. Short-term actions are favoured over long-term planning. In the off-the-cuff words of one of the global statesmen during CITES' existence: 'the truth is, no country is going to cut its growth or consumption substantially in the light of a long-term environmental problem'.[94] Until this attitude changes, CITES will remain as vital for the future of our world as it has been for the first 50 years of its existence.

NOTES

1 Challender, D.W.S., 't Sas-Rolfes, M., Broad, S. and Milner-Gulland, E.J. (2025) A Theory of Change to Improve Conservation Outcomes Through CITES. *Frontiers in Ecology and Evolution*, 13: 1425267. https://doi.org/10.3389/fevo.2025.1425267.
2 Resolution Conf. 9.24.
3 Challender, D.W.S., Cremona, P.J., Malsch, K., Robinson, J.E., Pavitt, A.T., Scott, J., Hoffman, R., Joolia, A., Oldfield, T.E.E, Jenkins, R.K.B., Conde, D.A., Hilton-Taylor, C. and Hoffmann, M. (2023) Identifying Species Likely Threatened by International Trade on the IUCN Red List Can Inform CITES Trade Measures. *Nature Ecology & Evolution*. https://doi.org/10.1038/s41559-023-02115-8.
4 Doc. 3.29.
5 Resolution Conf. 12.11.

6 Thomas, P. (2005) Mushrooms and the Future of CITES. *Endangered Species Bulletin*, 30 (2): 24–25.

7 Scanlon, J. (2023) *Forgotten Kingdom of Fungi!* https://illuminem.com/illuminem-voices/forgotten-kingdom-of-fungi (accessed 25.02.25).

8 Jenkins, M., Timoshyna, A. and Cornthwaite, M. (2018) *Wild at Home: Exploring the Global Harvest, Trade and Use of Wild Plant Ingredients*. TRAFFIC International, Cambridge and Schindler, C., Heral, E., Drinkwater, E., Timoshyna, A., Muir, G., Walter, S., Leaman, D.J. and Schippmann, U. (2022) *Wild Check—Assessing Risks and Opportunities of Trade in Wild Plant Ingredients*. FAO, Rome. https://doi.org/10.4060/cb9267en.

9 BirdLife International (2025) *Species Factsheet: Spix's Macaw Cyanopsitta spixii*. https://datazone.birdlife.org/species/factsheet/spixs-macaw-cyanopsitta-spixii (accessed 25.02.25).

10 SC78 Doc. 64.

11 CoP18 Doc. 83.1.

12 AC32 Doc. 14.2 Annex 2.

13 Downes, A.T. (2015) There's No Such Thing as a Sustainable Wildlife Trade. *The Dodo, Vox Media, LLC*. www.thedodo.com/attempts-to-manage-wildlife-tr-1001792311.html (accessed 25.02.25).

14 AC25 Inf. 1.

15 Martin, R. (2016) *CITES II or the Second Convention on International Trade in Endangered Species of Wild Fauna and Flora*. Available at SSRN: https://ssrn.com/abstract=2861388 or http://doi.org/10.2139/ssrn.2861388 https://papers.ssrn.com/sol3/Delivery.cfm/SSRN_ID2861388_code2595360.pdf?abstractid=2861388&mirid=1 (accessed 25.02.25).

16 SC31 Summary Report.

17 CoP9 Decision No. 1.

18 Notification to the Parties No. 951 of 29 January 1997.

19 Doc. SC.37.6.

20 Doc. 10.20.

21 Plen. 10.3 (Rev.).

22 Doc. Com. 10.26.

23 Dec. 10.111.

24 Dec. 10.59.

25 Dec. 11.1.

26 Resolution Conf. 14.2.

27 Resolution Conf. 18.3.

28 SC66 Doc. 30.2 Annex 2.

29 CoP19 Com. II. 2.

30 SC78 Summary Record.

31 CoP18 Doc. 11.

32 CoP18 Com. II Rec. 2 (Rev. 1).

33 SC74 Summary Record.

34 Doc. 5.38.

35 CoP16 Com I. Rec. 6 (Rev. 1).

36 Anon (2025) *Conservation Influencers*. IWMC—World Conservation Trust. www.iwmc.org/conservation-influencers/ (accessed 25.02.25).

37 Challender, D.W.S. and MacMillan, D.C. (2019) Investigating the Influence of Non-state Actors on Amendments to the CITES Appendices. *Journal of International Wildlife Law & Policy*, 22 (2): 90–114. https://doi.org/10.1080/13880292.2019.1638549.

38 Oberthur, S., Buck, M., Pfahl, S., Palmer, A., Werksman, J., Müller, S. and Tarasofsky, R. (2003) *Participation of Non-Governmental Organisations in International Environmental Governance: Legal Basis and Practical Experience*. Erich Schmidt Verlag, Berlin, Germany.

39 Challender, D.W.S. and MacMillan, D.C. (2019) Investigating the Influence of Non-state Actors on Amendments to the CITES Appendices. *Journal of International Wildlife Law & Policy*, 22 (2): 90–114. https://doi.org/10.1080/13880292.2019.1638549.

40 SC69 Doc. 11.3.

41 Resolution Conf. 17.3.

42 CoP18 Inf. 67 (Rev. 1) and CoP19 Inf. 100 (Rev. 1).

43 Notification to the Parties No. 2025/005 of 13 January 2025.

44 SC57 Doc. 13.2.

45 Anglada, M., Clement, S., Schraffl, F. and Zimmermann, M. (2009) *Green Meeting Guide*. United Nations Environment Programme. Freiburg, Germany and Paris, France.

46 CoP19 Doc. 10 Add Annex 1.

47 SC66 Inf.1.

48 Resolution Conf. 8.3.

49 CoP13 Com. I Rep. 10 (Rev. 1).

50 CITES Secretariat (2025) *CITES and Livelihoods*. https://cites.org/eng/prog/livelihoods (accessed 25.02.25).

51 Resolution Conf. 16.6 (Rev. CoP18).

52 CoP17 Doc. 13 and CoP18 Doc. 17.3.

53 CoP18 Com. II Rec. 4 (Rev. 1).

54 Doc. 10.59.

55 Resolution Conf. 13.10.

56 CoP14 Doc. 8.4.

57 Resolution Conf. 13.10 (Rev. CoP14).

58 Decs 15.15–15.17.

59 CoP16 Doc. 27 (Rev. 1).

60 AC26 Summary Record.

61 SC62 Summary Record.

62 Wildlife Conservation Society (2020) *Wildlife Conservation Society Statement in Support of Preventing Future Pandemics Act of 2020*. https://newsroom.wcs.org/News-Releases/articleType/ArticleView/articleId/14985/Wildlife-Conservation-Society-Statement-in-Support-of-Preventing-Future-Pandemics-Act-of–2020.aspx (accessed 25.02.25).

63 Ashe, D. and Scanlon, J.E. (2020) A Crucial Step Toward Preventing Wildlife-Related Pandemics. *Scientific American*, 15 June 2020 and Weissgold, B.J., Knights, P., Lieberman, S. and Mittermeier, R. (2020) *How We Can Use the CITES Wildlife Trade Agreement to Help Prevent Pandemics*. https://www.scientificamerican.com/article/how-we-can-use-the-cites-wildlife-tradeagreement-to-help-prevent-pandemics/ (accessed 25.02.25).

64 Decs 19.15 to 19.19.

65 CITES Secretariat (2024) *Memorandum of Understanding between the World Organisation for Animal Health and the Secretariat of the Convention on International Trade in Endangered Species of Wild Fauna and Flora*. https://cites.org/sites/default/files/eng/disc/coop/CITES_WOAH_MoU_final%20version_Feb2024.pdf (accessed 25.02.25).

66 CoP17 Doc. 20.

67 Resolution Conf. 17.5.

68 Anon (2025) *CITES Global Youth Network: The youth arm of The Convention on International Trade in Endangered Species of Wild Fauna and Flora (CITES)*. LinkedIn Corporation.   www.linkedin.com/company/cites-global-youth-network/   (accessed 25.02.25).

69 CoP19 Com. II Rec. 14 (Rev. 1).

70 Resolution Conf. 19.3.

71 Harris, K., Beintema, N., and Rosen, T. (Eds.) (2025) Summary of the 78th Meeting of the CITES Standing Committee: 3–8 February 2025. *Earth Negotiations Bulletin*, 21 (114). https://enb.iisd.org/cites-standing-committee-78-summary#brief-analysis-sc78 (accessed 25.02.25).

72 't Sas-Rolfes, M., Challender, D.W.S. and Wainwright, L. (2024) Playing the CITES Game: Lessons on Global Conservation Governance from African Megafauna. *Environmental Policy and Governance*: 1–16. https://doi.org/10.1002/eet.2123.

73 Lyman, E. (2023) It Takes Two: CITES, Illegal Wildlife Trade, and Importing Country Accountability. *William & Mary Environmental Law and Policy Review*, 47 (3): 707–748.

74 Dec. 19.10.

75 CoP17 Doc. 23.

76 Dec. 18.69.

77 CoP18 Doc. 28.

78 CITES Secretariat (2022) *World Wildlife Trade Report*. Geneva, Switzerland.

79 United Nations Office on Drugs and Crime (2024) *World Wildlife Crime Report 2024: Trafficking in Protected Species*. United Nations Publications. Vienna, Austria.

80 CITES Secretariat (2022) *World Wildlife Trade Report*. Geneva, Switzerland.

81 Nellemann, C. (Editor in Chief), Henriksen, R., Kreilhuber, A., Stewart, D., Kotsovou, M., Raxter, P., Mrema, E. and Barrat, S. (Eds.) 2016. *The Rise of Environ Mental Crime—A Growing Threat to Natural Resources Peace, Development and Security*. UNEP-INTERPOL Rapid Response Assessment. United Nations Environment Programme, Nairobi, Kenya.

82 INTERPOL (2023) *Illegal Wildlife Trade Has Become One of the 'World's Largest Criminal Activities*. www.interpol.int/en/News-and-Events/News/2023/Illegal-wildlife-trade-has-become-one-of-the-world-s-largest-criminal-activities (accessed 25.02.25).

83 United Nations Population Division (2025) https://population.un.org (accessed 25.02.25).

84 World Trade Organization (2025) *Evolution of Trade Under the WTO: Handy Statistics*. www.wto.org/english/res_e/statis_e/trade_evolution_e/evolution_trade_wto_e.htm (accessed 25.02.25).

85 World Bank Group (2025) *GDP Per Capita (Constant 2015 US$)—China*. https://data.worldbank.org/indicator/NY.GDP.PCAP.KD?end=2023&locations=CN&start=1975 (accessed 25.02.25).

86 Intergovernmental Science-Policy Platform on Biodiversity and Ecosystem Services (2019) *Global Assessment Report on Biodiversity and Ecosystem Services of the Intergovernmental Science-Policy Platform on Biodiversity and Ecosystem Services*. IPBES Secretariat, Bonn, Germany. https://doi.org/10.5281/zenodo.3831673.

87 CoP17 Doc. 27.

88 Chen, F. and 't Sas-Rolfes, M. (2021) Theoretical Analysis of a Simple Permit System for Selling Synthetic Wildlife Goods. *Ecological Economics*, 180 (106873). https://doi.org/10.1016/j.ecolecon.2020.106873.

89 Resolution Conf. 13.9.

90 CoP14 Doc. 48 (Rev. 1).

91 CoP14 Com. II Rep. 10 (Rev. 1).

92 International Maritime Organization (2024) *Revised Guidelines for the Prevention and Suppression of the Smuggling of Wildlife on Ships Engaged in International Maritime Traffic.* Convention on Facilitation of International Maritime Traffic Resolution 48/20/ Add.1 Annex 5. wwwcdn.imo.org/localresources/en/OurWork/Facilitation/Facilitation/ FAL.17%20(48).pdf (accessed 25.02.25).

93 Layne, E. (1973) Eighty Nations Write Magna Charta for Wildlife. *Audubon Magazine,* 75 (3): 99–102.

94 Blair, T. (2005) *Clinton Global Initiative Opening Plenary.* New York, 15 September. www.c-span.org/program/public-affairs-event/clinton-global-initiative-opening-ple-nary/161983 (accessed 25.02.25).

Annex 1
Selected Books About CITES

Inskipp, T. and Wells, S. (1979) *International Trade in Wildlife*. Routledge, London.

Emonds, G. (1981) *Guidelines for National Implementation of the Convention on International Trade in Endangered Species of Wild Fauna and Flora*. IUCN, Gland, Switzerland.

Favre, D.S. (1989) *International Trade in Endangered Species—A Guide to CITES*. M. Nijhoff Publ., Dordrecht, the Netherlands.

Fitzgerald, S. (1989) *International Wildlife Trade: Whose Business Is It*? World Wildlife Fund, Washington, DC.

Wijnstekers, W. (1990) *The Evolution of CITES: A Reference to the Convention on International Trade in Endangered Species of Wild Fauna and Flora*. CITES Secretariat, Switzerland. [there were 11 further editions of the book, with the last in 2018].

de Klemm, C. (1993) *Guidelines for Legislation to Implement CITES*. IUCN Environmental Policy and Law Paper No. 26 IUCN—The World Conservation Union, Gland, Switzerland.

Hemley, G. (Ed) (1994) *International Wildlife Trade: A CITES Sourcebook*. Island Press, Washington, DC.

Hutton, J. and Dickson, B. (Eds) (2000) *Endangered Species Threatened Convention: The Past, Present and Future of CITES, the Convention on International Trade in Endangered Species of Wild Fauna and Flora*. Taylor & Francis Group.

Reeve, R. (2002) *Policing International Trade in Endangered Species: The CITES Treaty and Compliance*. Earthscan, London.

Oldfield, S. (Ed) (2003) *The Trade in Wildlife: Regulation for Conservation*. Earthscan Publications, London.

Bowman, M., Davies, P. and Redgwell, C. (2010) The Convention on International Trade in Endangered Species of Wild Fauna and Flora. In: *Lyster's International Wildlife Law*. Cambridge University Press, 483–534.

Schneider, J. L. (2012) *Sold into Extinction: The Global Trade in Endangered Species*. Praeger.

Orenstein, R. (2013) *Ivory, Horn and Blood: Behind the Elephant and Rhinoceros Poaching Crisis*. Firefly Books, Richmond Hill, Canada.

Liljeblad, J. (2014) *The Convention on International Trade of Endangered Species: Local Authority and International Policy*. Quid Pro Books, New Orleans.

Sellar, J. M. (2014) *The UN's Lone Ranger: Combatting International Wildlife Crime*. Whittles Publishing, Dunbeath.

Wyatt, T. (2021) *Is CITES Protecting Wildlife*? Jenny Stanford Publishing.

Cordonier Segger, M.-C. (2023) *CITES as a Tool for Sustainable Development*. Greenwich Medical Media.

Lapointe, E. (2023) *Wildlife Betrayed: Why Prohibition Is Bad for Conservation and Development*. IWMC-World Conservation Trust, Lausanne, Switzerland.

Annex 2
Milestones of CITES

Mar 1973	Text of the Convention agreed and opened for signature
Apr 1974	UNEP contract for IUCN to provide CITES Secretariat starts
July 1975	Convention entered into force
Nov 1976	First meeting of the Conference of the Parties
May 1978	First Secretary-General starts work
June 1979	First meeting of the Standing Committee
Oct 1981	CITES logo officially adopted
Nov 1984	Secretariat transferred from IUCN to UNEP
May 1985	Start of the Review of Significant Trade
Nov 1985	First trade suspension compliance recommendation agreed
Jul 1987	CITES intersessional committees restructured to their present form
Jul 1987	First Secretariat review of alleged infractions by Parties submitted
Nov 1988	First meetings of the Animals and Plants Committees
Aug 1989	100th Party acceded to the Convention
Oct 1989	African elephant transferred to Appendix I
Mar 1992	First dedicated resolution on national laws to implementation of the Convention
Apr 1993	First trade suspension compliance recommendation over unsustainable trade in Appendix II species (Review of Significant Trade)
Nov 1994	Fort Lauderdale listing criteria adopted
Nov 1994	Current format of 'Resolutions' and 'Decisions' of the CoP agreed
Jun 1996	First Checklist of CITES Species published
Jun 1997	Standing Committee given authority to recommended trade suspension compliance recommendations for inadequate national laws
Aug 1997	CITES website first registered
Feb 1999	First approved sale of ivory stocks
Sep 1999	First trade suspension compliance recommendation over inadequate national laws
Apr 2000	150th Party acceded to the Convention
Apr 2000	First shark species to be added to the Appendices
Apr 2000	Standing Committee given authority to recommended trade suspension compliance recommendations for failure to submit annual trade reports
Dec 2002	First trade suspension complaince recommendation for failing to submit annual reports
June 2007	Guide to CITES compliance procedures agreed
Nov 2008	Second approved sale of ivory stocks
Nov 2010	International Consortium on Combating Wildlife Crime established

Oct 2016   Review of trade in animal specimens reported as produced in captivity
               begun
May 2021   First online CITES committee meeting (Standing Committee)
Oct 2023   CITES Trade database exceeds 25 million records
Jan 2025   185th Party acceded to the Convention
July 2025   50th anniversary of the entry into force of the Convention

Index

Note: page numbers in *italics* indicate a figure and page numbers in **bold** indicate a table on the corresponding page. Page numbers followed by 'n' with numbers refer to notes.